Laboratory Experiments in Microbiology

SIXTH EDITION

Ted R. Johnson
St. Olaf College

Christine L. Case
Skyline College

Benjamin
Cummings

An Imprint of Addison Wesley Longman

San Francisco • Boston • New York
Capetown • Hong Kong • London • Madrid • Mexico City
Montreal • Munich • Paris • Singapore • Sydney • Tokyo • Toronto

Publisher: Daryl Fox
Sponsoring Editor: Amy Folsom
Assistant Editor: Erin Joyce
Senior Production Editor: Larry Olsen
Production Service: Brian Jones
Design: Richard Kharibian
Cover Design: Yvo Riezebos
Composition: Thompson Type
Manufacturing Coordinator: Vivian McDougal

Photo Acknowledgments

Part One courtesy of Christine Case. Figure 1.1 courtesy of Leica. Part Two a&b courtesy of Christine Case. Part Eleven J. G. Hadley; Batelle-Pacific Northwest Laboratories/Biological Photo Service. Figure 44.1c, Christine Case. Part Twelve Dwayne C. Savage, University of Illinois at Urbana-Champaign. Color plates by Christine Case unless otherwise noted.

The authors and publisher believe that the lab experiments described in this publication, when conducted in conformity with the safety precautions described herein and according to the school's laboratory safety procedures, are reasonably safe for the students for whom this manual is directed. Nonetheless, many of the described experiments are accompanied by some degree of risk, including human error, the failure or misuse of laboratory or electrical equipment, mismeasurement, spills of chemicals, and exposure to sharp objects, heat, body fluids, blood, or other biologics. The authors and publisher disclaim any liability arising from such risks in connection with any of the experiments contained in this manual. If students have questions or problems with materials, procedures, or instructions on any experiment, they should always ask their instructor for help before proceeding.

ISBN 0-8053-7589-9
1 2 3 4 5 6 7 8 9 10—CRS—05 04 03 02 01 00

Addison Wesley Longman, Inc.
1301 Sansome Street
San Francisco, California 94111

Contents

Preface

Laboratory Experiments in Microbiology, Sixth Edition, is a laboratory companion to *Microbiology: An Introduction,* Seventh Edition, by Tortora, Funke, and Case. This edition includes updated exercises that reflect recent developments and current techniques in microbial genetics and immunology. Our goal is to provide a manual of basic microbiological techniques with applications for undergraduate students in diverse areas, including the biological sciences, the allied health sciences, agriculture, environmental science, nutrition, pharmacy, and various preprofessional programs. This manual contains 57 thoroughly class-tested exercises covering every area of microbiology. Most exercises require about one and one-half hours of laboratory time. By selecting an appropriate combination of exercises, the instructor can provide learning experiences to meet the needs of a particular course.

Our Approach

Our goal in writing this manual has been twofold—to teach microbiological techniques and to show students the importance of microbes in our daily lives and their central roles in nature. *All* of the laboratory techniques, thinking skills, and safety procedures recommended by the American Society for Microbiology are included in this manual.

Laboratory safety is our primary concern. Many students are preparing for work in a clinical environment and need to learn to handle biologically contaminated materials. Students preparing for work in biotechnology and other laboratories must master aseptic techniques to avoid contamination of their samples. Students must not only learn but also practice these safety techniques so that safety becomes part of their professional behavior. We have followed the Centers for Disease Control and Prevention guidelines for the safe handling of microbes and human body fluids. Students are instructed to work only with their own body fluids. See the section entitled Safety in the Laboratory on page 2 and the following sections of the Introduction for specific safety suggestions. At key points in the exercises appear safety boxes, marked with either a biohazard logo or a general safety logo, indicating appropriate safety techniques.

Almost every exercise includes an actual experiment requiring students to analyze data. We hope in this way to promote *analytical reasoning* and to make laboratory sessions interesting for students as well as to provide a variety of opportunities for *reinforcement* of the technical skills they have learned. Each exercise has a series of Critical Thinking questions to enhance students' investigative skills. To demonstrate further some practical uses of microbiology, we have frequently included material with direct application to procedures performed in clinical and commercial laboratories.

Scope and Sequence

This manual is divided into 13 parts. Each part introduction explains the unifying theme for that part. Exercises in the first three parts provide sequential development of fundamental techniques. The remaining exercises are as independent as possible to allow the instructor to select the most desirable sequence. A Techniques Required section preceding each experiment lists prerequisites from earlier exercises.

Part One, Microscopy, emphasizes observation through the microscope. Practice in use and care of the microscope is followed by observations of microbes, to familiarize students with their sizes. Phase contrast and brightfield microscopy are used to examine living material.

Part Two, Staining Methods, begins with an explanation of how to handle bacterial cultures and teaches the use of the most common stains, including preparation of stained samples and their examination. A highly reliable capsule stain is included. Exercise 8, Morphologic Unknown, introduces the concept of unknowns and can be used to test knowledge of staining methods.

Part Three, Cultivation of Bacteria, stresses aseptic technique and covers the isolation of bacterial strains and the maintenance of bacterial cultures. An exercise dealing with special media prepares the student for the next part.

Part Four, Microbial Metabolism, includes five exercises on bacterial metabolism that provide the tools for Exercises 18, 33, and 51, Unknown Identifications. These exercises are especially useful for learning biochemistry. Exercise 18, Rapid Identification Methods, demonstrates computerization of laboratory work.

Part Five, Microbial Growth, deals with the effects of environmental conditions, such as temperature and the presence of oxygen, on growth.

Part Six, Control of Microbial Growth, provides practical applications of concepts of microbial growth. Methods of controlling unwanted microbes in food or in a clinical environment are examined in this part.

Part Seven, Microbial Genetics, is of particular interest because of recent notable advances in this field. Exercise 27 is a new experiment in which students look at expression in two operons using Green Fluorescent Protein. Exercise 28 demonstrates the isolation of bacterial mutants. Exercise 29 allows students to isolate DNA and transform cells without special equipment. DNA fingerprinting is introduced in Exercise 30. Transformation by an antibiotic resistance plasmid is performed in Exercise 31. Restriction enzyme digestion and agarose gel electrophoresis are used to analyze the plasmid. In Exercise 32, suspected chemical carcinogens are tested by the Ames Test.

Part Eight, The Microbial World, examines the diversity of microorganisms. Information and techniques learned in previous exercises are used to identify a bacterial unknown in Exercise 33. Cyanobacteria are included in Exercise 36, Phototrophs. The morphology and ecological niches of free-living eukaryotic organisms studied in microbiology are examined in Exercises 34 through 37.

Part Nine, Viruses, provides an opportunity to isolate, cultivate, and quantify bacteriophages. Students determine the host range of a plant virus in Exercise 39.

Part Ten, Interaction of Microbe and Host, introduces basic concepts of epidemiology. Methods of tracking and identifying causes of infectious diseases are practiced in the exercises in this part.

Part Eleven, Immunology, covers the host's response to infectious disease with exercises on nonspecific resistance and agglutination tests, and a new exercise using the ELISA technique.

Part Twelve, Microorganisms and Disease, emphasizes procedures employed in the clinical laboratory. Hands-on exercises on normal and pathogenic bacteria in the human body are supplemented with examples from the Bacteria ID CD-ROM to allow students to practice identifying bacteria. In Exercise 51, students identify an unknown from a simulated clinical sample.

Part Thirteen, Microbiology and the Environment, includes standard methods for the examination of food. The MUG test is used to examine water for microbiological quality. Exercise 55, Microbes Used in the Production of Foods, offers an opportunity to study food microbiology. Exercise 56, Microbes in Soil: The Nitrogen and Sulfur Cycles, provides an example of biogeochemical cycles and soil microbiology. Exercise 57, Microbes in Soil: Bioremediation, is a unique exercise illustrating the use of bacteria for cleaning up environmental pollution.

The appendices at the end of the manual provide a convenient reference to techniques required in several exercises. Appendix H includes several identification keys to bacteria.

Organization of Each Exercise

The exercises in this manual may involve mastering a skill or procedure or understanding a particular concept. Most of the exercises are investigative by design, and the student is asked to analyze the experimental results and draw conclusions. The exercises are organized as follows:

- Each exercise begins with a section called *Objectives,* which lists skills or concepts to be mastered in that exercise. The objectives can be used to test mastery of the new material after completing the exercise.

- The *Background* provides definitions and explanations for each exercise. The student is expected to refer to a textbook for more detailed explanations of the concepts introduced in the laboratory exercises.

- *Materials* lists include supplies, media, and equipment needed for the exercise.

- *Cultures* lists the living organisms required for the exercise.

- *Techniques Required* gives a list of techniques needed to complete each exercise.

- *Procedure* then provides step-by-step instructions, stated as simply as possible and frequently supplemented with diagrams. Questions are occasionally asked in the Procedure section to remind the student of the rationale for a step.

- The *Laboratory Report* is designed to help students learn to collect and present data in a systematic fashion. Students are asked to write the *Purpose* of the exercise so they can relate the laboratory activity to their learning. Where appropriate, exercises ask students to formulate their *Hypothesis.* Tables are provided to record *Data* or *Results.* Occasionally, students are asked to write their *Expected Results* using information provided in the Background and their own experience. Students may be asked to write their *Conclusions* following the Data or Results. Usually, *Questions* in the Laboratory Report ask for interpretation of results. The Questions are designed to lead the student from a collection of data or observations to a conclusion. In most instances, the results for each student team will be unique; they can be compared with the information given in the Background and other references but will not be identical to those references. The range of questions in each exercise requires students to think about their results, recall facts, and then use this information to answer the questions. *Critical Thinking* questions are designed to help students use their new knowledge and practice analytic skills.

Illustrations

This book is generously illustrated with diagrams of procedures and 12 pages of photographs in full color. There is a photo quiz in the color section in which students identify an unknown.

For additional photographs of lab results, students can visit the laboratory supplement on The Microbiology Place Web site (http://www.microbiologyplace.com). Photographs of media and techniques help students interpret their results without giving answers and actual lab results.

Preparation Guide

The comprehensive *Preparation Guide* (ISBN 0-8053-7586-4) provides the instructor with all the information needed to set up and teach a laboratory course with this manual. It includes the following:

- General instructions for setting up the lab.
- Information on obtaining and preparing cultures, media, and reagents.
- A master table showing the techniques required for each exercise.
- Cross-references for each exercise to specific pages in *Microbiology: An Introduction,* Seventh Edition.
- For each exercise: helpful suggestions, detailed lists of materials needed, and answers to all the questions in the student manual.

To make *Laboratory Experiments in Microbiology,* Sixth Edition, easy to use, we have carefully designed the experiments to use inexpensive, readily available, nonhazardous materials. Moreover, the exercises have been thoroughly tested in our classes in California and Minnesota by students with a wide variety of talents and interests. Our students have enjoyed their microbiology laboratory experiences; we hope yours will, too!

Acknowledgments

We are most grateful to the following individuals for their time, talent, and interest in our work. Each person carefully read and edited critical parts of the manuscript.

Chuck Hoover of the University of California, San Francisco, for making us aware of the new techniques used in dental microbiology for Exercise 48.

Nick Kapp of Skyline College for providing a novel transformation experiment for Exercise 29.

Anne Jayne of the University of San Francisco for offering suggestions and a capsule stain.

Patricia Carter of Skyline College for her invaluable assistance in preparing materials and proofreading.

We are indebted to St. Olaf College and Skyline College for providing the facilities and resources in which innovative laboratory exercises can be developed.

We would like to commend the staff at Benjamin Cummings for their support. In particular, we thank Amy Folsom, our Sponsoring Editor; Brian Jones, for his editorial skill and meticulous care; and Larry Olsen, for expertly guiding this manual through the production process.

Last, but not least, our gratitude goes to Don Biederman, who provided timely encouragement and support, and Michelle Johnson, who gave her professional insights and was a sustaining presence.

Reviewers

John Curtis Branch, Oklahoma City College

Judith K. Davis, Florida Community College at Jacksonville

Michael Griffin, Angelo State University

Catherine O'Brien, San Jacinto College South

Harriet Solms Mathews, Pasco Hernando Community College

Curt W. Spanis, University of San Diego

A Special Note to Students

This book is for you. The study of microbiology is dynamic because of the diversity of microbes and the variability inherent in every living organism. Outside of the laboratory—on a forest walk or tasting a fine cheese—we experience the activities of microbes. We want to share our excitement for studying these small organisms. Enjoy!

Ted R. Johnson
Christine L. Case

Introduction

*Life would not long remain possible in the
absence of microbes.*

LOUIS PASTEUR

Welcome to microbiology! Microorganisms are all around us, and as Pasteur pointed out a century ago, they play vital roles in the ecology of life on Earth. In addition, some microorganisms provide important commercial benefits through their use in the production of chemicals (including antibiotics) and certain foods. Microorganisms are also major tools in basic research in the biological sciences. Finally, as we all know, some microorganisms cause disease—in humans, other animals, and plants.

In this course, you will have firsthand experience with a variety of microorganisms. You will learn the techniques required to identify, study, and work with them. Before getting started, you will find it helpful to read through the suggestions on the next few pages.

Suggestions to Help You Begin

1. Science has a vocabulary of its own. New terms will be introduced in **boldface** throughout this manual. To develop a working vocabulary, make a list of these new terms and their definitions.
2. Because microbes are not visible without a microscope, common names have not been given to them. The word *microbe*, now in common use, was introduced in 1878 by Charles Sedillot. The microbes used in the exercises in this manual are referred to by their *scientific names*. The names will be unfamiliar at first, but do not let that deter you. Practice saying them aloud. Most scientific names are taken from Latin and Greek roots. If you become familiar with these roots, the names will be easier to remember.
3. Microbiology usually provides the first opportunity undergraduate students have to experiment with *living organisms*. Microbes are relatively easy to grow and lend themselves to experimentation. Since there is variability in any population of living organisms, not all the experiments will "work" as the lab manual says. The following exercise will illustrate what we mean:

Write a description of *Homo sapiens* for a visitor from another planet: _____
_____
_____
_____
_____
_____
_____

After you have finished, look around you. Do all your classmates fit the description exactly? Probably not. Moreover, the more detailed you make your description, the less conformity you will observe. During lab, you will make a very detailed description of an organism and probably find that this description does not match your reference exactly.

4. Microorganisms must be cultured or grown to complete most of the exercises in this manual. Cultures will be set up during one laboratory period and will be examined for growth in the next laboratory period. Accurate record keeping is therefore essential. Mark the steps in each exercise with a bright color or a bookmark, so you can return to complete your Laboratory Report on that exercise. *Accurate records* and *good organization* of laboratory work will enhance your enjoyment and facilitate your learning.
5. *Observing* and *recording* your results carefully are the most important parts of each exercise. Ask yourself the following questions for each experiment:

   What did the results indicate?

   Are they what I expected? If not, what happened?
6. If you do not master a technique, try it again. In most instances, you will need to use the technique again later in the course.
7. Be sure you can answer the questions that are asked in the Procedure for each exercise. These questions are included to reinforce important points that will ensure a successful experiment.
8. Finally, carefully study the general procedures and safety precautions that follow.

General Procedures in Microbiology

In many ways, working in a microbiology laboratory is like working in the kitchen. As some famous chefs have said:

> *Our years of teaching cookery have impressed upon us the fact that all too often a debutant cook will start in enthusiastically on a new dish without ever reading the recipe first. Suddenly an ingredient, or a process, or a time sequence will turn up, and there is astonishment, frustration, and even disaster. We therefore urge you, however much you have cooked, always to read the recipe first, even if the dish is familiar to you. . . . We have not given estimates for the time of preparation, as some people take half an hour to slice three pounds of mushrooms, while others take five minutes.**

1. Read the laboratory exercises *before* coming to class.
2. *Plan* your work so that all experiments will be completed during the assigned laboratory period. A good laboratory student, like a good cook, is one who can do more than one procedure at a time—that is, one who is efficient.
3. Use only the *required* amounts of materials, so that everyone can do the experiment.
4. *Label* all of your experiments with your name, date, and lab section.
5. Even though you will do most exercises with another student, you must become familiar with all parts of each exercise.
6. Keep *accurate* notes and records of your procedures and results, so that you can refer to them for future work and tests. Many experiments are set up during one laboratory period and observed for results in the next laboratory period. Your notes are essential to ensure that you perform all the necessary steps and observations.
7. *Demonstrations* will be included in some of the exercises. Study the demonstrations and learn the content.
8. Let your instructor know if you are color-blind; many techniques require discrimination of colors.
9. Keep your cultures current; discard old experiments.
10. *Clean up* your work area when you are finished. Leave the laboratory clean and organized for the next student. Remember:

    Stain and reagent bottles should be returned to their original location.

*J. Child, L. Bertholle, and S. Beck. *Mastering the Art of French Cooking,* vol. 1. New York: Knopf, 1961.

Slides should be washed and put back into the box clean.

All markings on glassware (e.g., Petri plates and test tubes) should be removed before putting glassware into the marked autoclave trays.

Glass Petri plates should be placed agar-side down in marked autoclave containers.

Swabs and pipettes should be placed in the appropriate disinfectant jars or biohazard containers.

Disposable plasticware should be placed in marked autoclave containers.

Used paper towels should be discarded.

Safety in the Laboratory

During your microbiology course, you will learn how to safely handle fluids containing microorganisms. Through practice you will be able to perform experiments so that bacteria and viruses remain in the desired containers, uncontaminated by microbes in the environment. These techniques, called **aseptic techniques,** will be a vital part of your work if you are going into health care or biotechnology.

1. Do not eat, drink, smoke, store food, or apply cosmetics in the laboratory.
2. Wear shoes at all times in the laboratory.
3. Disinfect work surfaces at the beginning and end of every lab period and after every spill. The disinfectant used in this laboratory is _____ .
4. Wash your hands after every laboratory exercise. Because bar soaps may become contaminated, use liquid or powdered soaps.
5. Use mechanical pipetting devices; do not use mouth pipetting.
6. Place a disinfectant-soaked paper towel on the desk while pipetting.
7. Wash your hands immediately and thoroughly if they are contaminated with microorganisms.
8. Cover spilled microbial cultures with paper towels, and disinfect the towels. Leave for 20 minutes, then clean up the spill.
9. Do not touch broken glassware with your hands; use a broom and dustpan. Place broken glassware contaminated with microbial cultures or body fluids in the To Be Autoclaved container. (See below, Specific Hazards, for what to do with broken glassware that is not contaminated.)
10. Place glassware and slides contaminated with blood, urine, and other body fluids in disinfectant.
11. Tie long hair back.
12. Don't perform unauthorized experiments.
13. Don't use equipment without instruction.
14. Don't engage in horseplay in the laboratory.

Procedures marked with this biohazard icon should be performed carefully, to minimize the risk of transmitting disease.

15. Work only with your own body fluids and wastes in exercises requiring saliva, urine, blood, or feces, to prevent transmission of disease. The Centers for Disease Control and Prevention (CDC) state that "epidemiologic evidence has implicated only blood, semen, vaginal secretions, and breast milk in transmission of HIV" (*MMWR* 36 [S2], 8/21/87).

16. If you got this far in the instructions, you'll probably do well in lab. Enjoy lab, and make a new friend.

Specific Hazards

 Procedures marked with this safety icon should be performed carefully, to minimize risk of exposure to chemicals or fire.

Alcohol
Keep containers of alcohol away from open flames.

Glassware Not Contaminated with Microbial Cultures
1. If you break a glass object, sweep up the pieces with a broom and dustpan. Do not pick up pieces of broken glass with your bare hands.
2. Place broken glass in one of the containers marked for this purpose. The one exception to this rule concerns broken thermometers; consult your instructor if you break a thermometer.

Electrical Equipment
1. The basic rule to follow is: Electricity and water don't mix. Do not allow water or any water-based solution to come into contact with electrical cords or electrical conductors. Make sure your hands are dry when you handle electrical connectors.
2. If your electrical equipment crackles, snaps, or begins to give off smoke, do not attempt to disconnect it. Call your instructor immediately.

Fire
1. If *gas burns* from a leak in the burner or tubing, turn off the gas.
2. If you have a *smoldering sleeve,* run water on the fabric.
3. If you have a *very small fire,* the best way to put it out is to smother it with a towel or book (not your hand). Smother the fire quickly.
4. If a *larger fire* occurs, such as in a wastebasket or sink, use one of the fire extinguishers in the lab to

put it out. Your instructor will demonstrate the use of the fire extinguishers.

5. In case of a *large fire* involving the lab itself, evacuate the room and building according to the following procedure:
   a. Turn off all gas burners, and unplug electrical equipment.
   b. Leave the room and proceed _____ _____ .
   c. It is imperative that you assemble in front of the building so that your instructor can take roll to determine if anyone is still inside. Do not wander off.

Accidents and First Aid
1. Report all accidents immediately. Your instructor will administer first aid as required.
2. For spills in or near the eyes, use the eyewash immediately.
3. For large spills on your body, use the safety shower.
4. For heat burns, chill the affected part with ice as soon as possible. Call your instructor.
5. Place a bandage on any cut or abrasion.

Power Outage
If the electricity goes off, be sure to turn your gas jet off. When the power is restored, the gas will come back on.

Earthquake
Turn off your gas jet and get under your lab desk during an earthquake. Your instructor will give any necessary evacuation instructions.

Orientation Walkabout
Locate the following items in the lab:

Broom and dustpan

Eyewash

Fire blanket

Fire extinguisher

First aid cabinet

Fume hood

Instructor's desk

Reference books

Safety shower

To Be Autoclaved area

Biohazard containers

Special Practices
1. Keep laboratory doors closed when experiments are in progress.
2. The instructor controls access to the laboratory and allows access only to people whose presence is required for program or support purposes.

3. Place contaminated materials that are to be decontaminated at a site away from the laboratory into a durable, leakproof container that is closed before being removed from the laboratory.
4. An insect and rodent control program is in effect.
5. A needle should not be bent, replaced in the sheath, or removed from the syringe following use. Place the needle and syringe promptly in a puncture-resistant container and decontaminate, preferably by autoclaving, before discarding them.
6. Inform your instructor if you are pregnant, are taking immunosuppressive drugs, or have any other medical condition (e.g., diabetes, immunologic defect) that might necessitate special precautions in the laboratory.
7. Potential pathogens used in the exercises in this manual are classified in Class 1 by the U.S. Public Health Service. These bacteria present minimal hazard and require ordinary aseptic handling conditions. No special competence or containment is required. These organisms are the following:

Corynebacterium species
Enterobacter species
Mycobacterium species
Proteus species
Pseudomonas aeruginosa
Salmonella species
Serratia marcescens
Shigella species
Staphylococcus species
Streptococcus species

Laboratory Facilities

1. Interior surfaces of walls, floors, and ceilings are water resistant so that they can be easily cleaned.
2. Benchtops are impervious to water and resistant to acids, alkalis, organic solvents, and moderate heat.
3. Windows in the laboratory are closed and sealed.
4. An autoclave for decontaminating laboratory wastes is available, preferably within the laboratory.

Contact with Blood and Other Body Fluids

The following procedures should be used by all health care workers, including students, whose activities involve contact with patients or with blood or other body fluids. While these procedures were developed by the CDC* to minimize the risk of transmitting HIV in a health care environment, adherence to these guidelines will minimize transmission of *all* nosocomial infections.

1. Wear gloves for touching blood and body fluids, mucous membranes, or nonintact skin and for handling items or surfaces soiled with blood or body fluids. Change gloves after contact with each patient.
2. Wash hands and other skin surfaces immediately and thoroughly if they are contaminated with blood or other body fluids.
3. Wear masks and protective eyewear or face shields during procedures that are likely to generate droplets of blood or other body fluids.
4. Wear gowns or aprons during procedures that are likely to generate splashes of blood or other body fluids.
5. Wash hands and other skin surfaces immediately after gloves are removed.
6. Mouthpieces, resuscitation bags, or other ventilation devices should be available for use in areas in which the need for resuscitation is predictable. Emergency mouth-to-mouth resuscitation should be minimized.
7. Health care workers who have exudative lesions or weeping dermatitis should refrain from all direct patient care and from handling patient care equipment.
8. Pregnant health care workers are not known to be at greater risk of contracting HIV infection than health care workers who are not pregnant; however, if a health care worker develops HIV infection during pregnancy, the infant is at risk of infection. Because of this risk, pregnant health care workers should be especially familiar with and strictly adhere to precautions to minimize the risk of HIV transmission.
9. In a laboratory exercise where human blood is used, students should wear gloves or work only with their own blood and should dispose of all slides and blood-contaminated materials immediately after use. Any cuts or scrapes on the skin should be covered with a sterile bandage.

*CDC. "Recommendations for Prevention of HIV Transmission in Health-Care Settings." *MMWR* 38 (S2), 1989.

I.1　Gram-negative rods.
Escherichia coli (Exercise 5).

I.2　Gram-positive cocci. *Staphylococcus aureus* (Exercise 5).

I.3　Flagella stain. Peritrichous flagella of *Proteus vulgaris* (Exercise 7).

I.4　Negative stain.
Staphylococcus aureus (Exercise 4).

II.1　Ames Test. The chemical (ethidium bromide) diffused from the disk on the left and caused the his⁻ bacteria to revert to his⁺. The his⁻ cells on the control plate (right) are unable to grow on glucose-minimal salts agar (Exercise 32).

II.2　DNA is collected on a glass rod (Exercise 29).

II.3　Restriction enzyme digests of small pieces of DNA. To see the bands, the DNA was stained with ethidium bromide and the gel was illuminated with UV light (Exercise 30).

II.4　*E. coli*. The colonies fluoresce under UV light because the cells contain the jellyfish gene for Green Fluorescent Protein (Exercise 27).

PLATE III

III.1   Reactions in OF-glucose medium. The species in tubes **1** and **2** is an oxidizer. The organism in tubes **3** and **4** does not use glucose. The culture in tubes **5** and **6** is a fermenter (Exercise 13).

III.2   Reactions in fermentation tubes. 1 is an uninoculated control. Growth and acid production from carbohydrate fermentation are seen in **2**. Acid and gas are produced from fermentation in **3** (Exercise 14).

III.3   Reactions in litmus milk. 1 is an uninoculated control. **2** shows an alkaline reaction. Acid is produced in **3**. The litmus is reduced in **4**. The clear fluid layer in **5** is the result of peptonization (Exercise 15).

III.4   Peptone iron agar is used to detect the production of H₂S. H_2S produced in the tube will precipitate as ferrous sulfide (Exercise 16).

III.5   Twenty biochemical tests are performed in an API 20E® strip. The top strip is inoculated with *Proteus vulgaris* and the bottom strip with *Escherichia coli* (Exercise 18).

III.6   Gelatin hydrolysis. After hydrolysis (**1**), gelatin remains liquid. **2** is unhydrolyzed gelatin (Exercise 15).

III.7   Fifteen biochemical tests are performed in Enterotube® II. The bottom tube is uninoculated (Exercise 18).

MICROBIAL METABOLISM

III.8   Methyl red test. Red color (in **1**) after addition of methyl red indicates a positive test. **2** is methyl red–negative (Exercise 14).

III.9   Voges–Proskauer test. A positive Voges–Proskauer test develops a red color when exposed to oxygen (Exercise 14).

III.10   Citrate test. Utilization of citric acid as the sole carbon source in Simmons citrate agar causes the indicator to turn blue (**2**). **1** is citrate-negative (Exercise 18).

III.11   Oxidase test. Colonies of cytochrome oxidase–positive bacteria turn black as oxidase reagent diffuses from the oxidase test disk into the nutrient agar (Exercise 17).

III.12   Reactions in triple sugar iron (TSI) agar. 1 shows growth with no fermentation or hydrogen sulfide (H_2S) production. **2** shows blackening due to H_2S and acid and gas from fermentation of glucose and sucrose and/or lactose. Sucrose and lactose were not fermented in **3**; H_2S production masks the glucose fermentation reaction although gas is produced. Acid and gas are produced from glucose in **4**. **5** is uninoculated. **6** shows acid and gas production from glucose and sucrose and/or lactose (Exercise 49).

III.13   Urease production. Hydrolysis of urea produces ammonia, which turns the indicator fuchsia (Exercise 15).

III.14   MIO medium. The culture in the tube on the left is *motile*. The red color of the Kovacs reagent added over the agar indicates *indole* production in the middle tube. Removal of CO_2 from *ornithine* turns the indicator purple, as seen in the tube on the right (Exercise 16).

III.15   Phenylalanine deaminase. Removal of the amino group produces an organic acid that forms a green complex with the ferric ion–containing indicator (Exercise 16).

PLATE IV   **EUKARYOTES**

IV.1   The yeast *Saccharomyces cerevisiae* produces circular, convex, glistening colonies on Sabouraud agar (Exercise 34). The plate on the right was irradiated for 25 seconds (Exercise 23).

100 µm

IV.2   *Rhizopus nigricans* growing on styrofoam impregnated with a nutrient medium. Dark sporangia are visible at the tips of the sporangiophores (Exercise 35).

IV.3   *Penicillium* has a white mycelium and green conidiospores. Note the crystallized penicillin on top of the mold colony (Exercise 35).

100 µm

IV.4   Red snow at Tioga Pass (Yosemite National Park) due to the green alga *Chlamydomonas nivalis.* The green color of chlorophyll is masked by red carotenoid pigments produced in the presence of intense light (Exercise 36).

IV.5   The green alga *Spirogyra* and protozoan *Paramecium*, in pond water (Exercises 36 and 37).

V.2 ***Staphylococcus aureus* growing on mannitol salt agar.** The yellow color indicates that mannitol is fermented. The red plate is uninoculated (Exercise 46).

V.1 ***Serratia marcescens* colonies on nutrient agar after incubation at 25°C.** Note color variations due to mutations in prodigiosin production (Exercise 28).

V.3 ***Pseudomonas aeruginosa* produces a water-soluble blue pigment on Pseudomonas agar P** (Exercise 50).

V.4 Water-soluble pigment of *Pseudomonas aeruginosa* fluoresces under ultraviolet light (Exercise 50).

V.5 Starch hydrolysis. Iodine is used to detect the presence of starch on a starch agar plate. The clearing around the *Bacillus* indicates that the starch was hydrolyzed by an extracellular enzyme (Exercise 13).

V.6 Compare the effects of dry heat and autoclaving. The control (left) is a pour plate inoculated with soil. The middle plate was inoculated with soil exposed to dry heat (121°C) for one hour. The soil on the right was autoclaved (121°C) and has no bacterial growth (Exercise 22).

V.7 Colonies of *Aspergillus niger* have white mycelia and black conidiospores (Exercise 35).

PLATE VI   ENVIRONMENTAL MICROBIOLOGY

VI.1   *Bacillus cereus* var. *mycoides* colonies. Easily recognized by their filamentous form. Bending of the filaments to the left or right is a strain characteristic (Exercise 11).

VI.2   Actinomycete colonies. *Streptomyces griseus* colonies penetrate into the agar as well as extend above. Powdery appearance in the corner is due to conidiospores (Exercise 11).

VI.3   Photosynthetic bacteria. Purple and green sulfur bacteria have colonized in this Winogradsky column (Exercise 56).

VI.4   Root nodules. *Rhizobium leguminosarum* infected the roots of this vetch plant and caused the production of pink nodules (Exercise 56).

VI.5   New antibiotics are discovered by examining colonies of soil bacteria. In this sample, one species is producing an antibiotic that inhibits *Bacillus cereus* from growing near it (Exercise 9).

VII.1   Numerous plaques (clearings) in this *Escherichia coli* culture are due to growth of a T-even bacteriophage (Exercise 38).

VII.2   Chlorosis (loss of green color) and plaques (spotting) in this tomato leaf are due to a tobacco mosaic virus infection (Exercise 39).

VII.3   Carrot soft rot caused by *Erwinia carotovora* (Exercise 41).

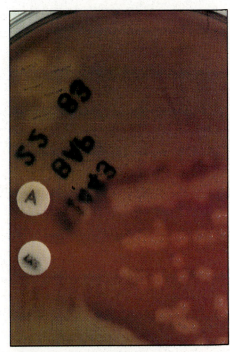

VII.5   Rapid identification of *Streptococcus pyogenes*. *S. pyogenes* is β-hemolytic and inhibited by Bacitracin (Exercise 47).

VII.4   Milky spore disease. The larva on the right ingested the toxin produced by *Bacillus thuringiensis* (Exercise 41).

PLATE VIII   # MEDICAL MICROBIOLOGY

VIII.1   Antibiotic sensitivity of this *Staphylococcus aureus* **culture is demonstrated by the Kirby–Bauer agar diffusion test** (Exercise 25).

VIII.2   Some of the cells of this strain of *Staphylococcus aureus* **are resistant to penicillin** (Exercise 25).

VIII.3   Coagulase test. Many pathogenic strains of *Staphylococcus aureus* can coagulate plasma. Bacteria are caught in fibrin clumps in **2** (Exercise 46).

VIII.4   Growth of *Streptococcus pneumoniae* **on blood agar demonstrates alpha-hemolysis.** Note the greenish color around the colonies (Exercise 47).

VIII.5   Beta-hemolysis produces a clear area around the colonies of *Streptococcus pyogenes* **on blood agar** (Exercise 47).

Anti-B   **Anti-A**

IX.1 Slide agglutination test for the ABO blood group. A negative reaction with anti-B (left) and a positive reaction with anti-A (indicated by agglutination) on the right (Exercise 43). What is this person's blood type?

X.1 Soil is excavated and layered with pipes at a former truck stop. The pipes are connected to supplies of nitrogen, phosphate, and oxygen to provide additional nutrients for bacteria that use hydrocarbons as their carbon source (Exercise 57). (Courtesy of Randall von Wedel, Cytoculture International.)

X.2 Bacteria that degrade the hydrocarbon in the agar produce water-soluble products, resulting in clearing around the colonies (Exercise 57).

X.3 In a wastewater treatment facility, sewage is aerated to encourage bacterial growth and decomposition of organic matter (Exercises 52 and 53).

X.4 After secondary sewage treatment, water is disinfected and discharged into a water course or onto land (Exercises 52 and 53).

PLATE XI   PLATE CULTURE FOR ENTERICS

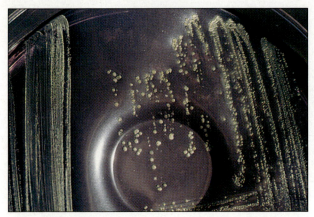

XI.1   Colonies of *Escherichia coli* (shown) and *Citrobacter* develop a metallic green sheen on EMB agar (Exercise 53).

XI.2   Colonies of *Enterobacter* have a distinctive blue "fish-eye" appearance on EMB agar (Exercise 53).

XI.3   Nonlactose-fermenters such as *Shigella boydii* (shown) produce colorless colonies on EMB agar (Exercise 53).

XI.5   *E. coli* is identified by its ability to convert MUG to a fluorescent compound that is visible with an ultraviolet lamp (Exercise 52).

XI.4   Coliforms trapped on a membrane filter grow into colonies when the filter is laid on EMB agar and incubated (Exercise 53).

XI.6   MacConkey agar is selective for gram-negative enterics. Colonies of lactose-fermenting enterics are red and surrounded by an opaque precipitate (left). Lactose-negative bacteria produce colorless colonies (right) (Exercises 49 and 50).

XI.7   Cells of *Proteus mirabilis* are highly motile. *P. mirabilis* colonies exhibit swarming over EMB agar as young cells move away from the colony's center (Exercises 50 and 53).

This is an example of how to identify an unknown bacterium.

XII.2   A Gram stain is made from the pure culture (Exercise 5).

XII.1   The unknown organism is isolated in pure culture (Exercise 11).

XII.3   Based on the result of the Gram stain, a catalase test is indicated (Exercise 17).

XII.4   Results of the previous tests lead to the next series of tests. A glucose fermentation tube is inoculated with the unknown bacterium (Exercise 13).

Using the identification keys in Appendix H of this manual, identify the genus and species of this bacterium.

XII.5   Lipid hydrolysis test. The light color around the colonies indicates hydrolysis of the lipid in the spirit blue medium (Exercise 57).

Answer: *Micrococcus luteus*

PLATE XIII

PHOTOTROPHIC MICROORGANISMS

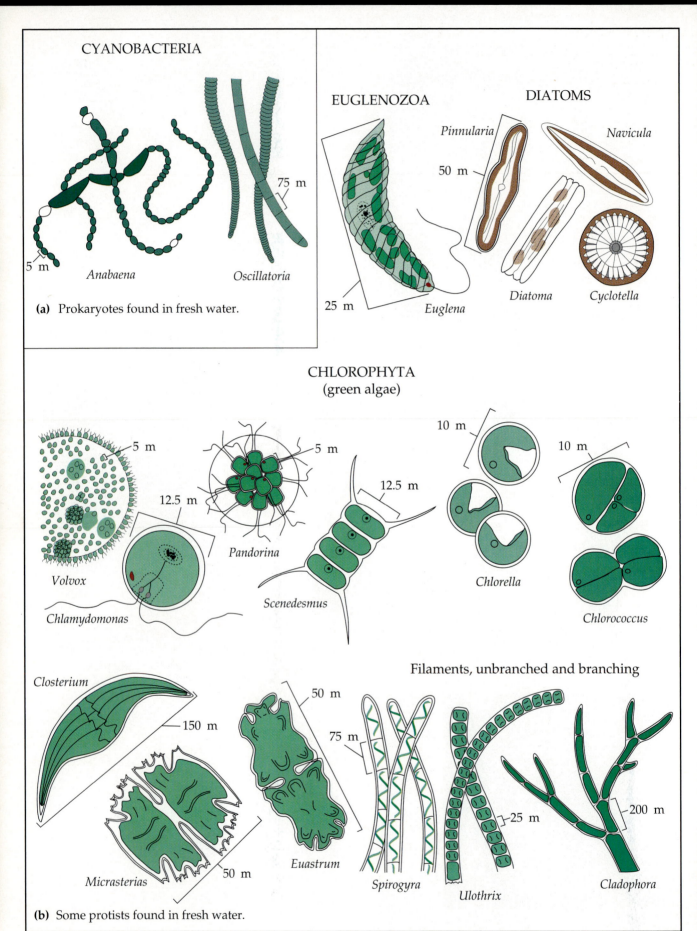

CYANOBACTERIA

5 m
Anabaena

75 m
Oscillatoria

(a) Prokaryotes found in fresh water.

EUGLENOZOA

25 m
Euglena

DIATOMS

Pinnularia

50 m

Navicula

Diatoma

Cyclotella

CHLOROPHYTA
(green algae)

5 m
Volvox

12.5 m
Chlamydomonas

5 m
Pandorina

12.5 m
Scenedesmus

10 m
Chlorella

10 m
Chlorococcus

Closterium

150 m

Micrasterias

50 m

Euastrum

50 m

Filaments, unbranched and branching

75 m
Spirogyra

25 m
Ulothrix

200 m
Cladophora

(b) Some protists found in fresh water.

part one

Microscopy

EXERCISES

1 Use and Care of the Microscope
2 Examination of Living Microorganisms

The microscope is a very important tool for a microbiologist. Microscopes and microscopy (microscope technique) are introduced in Exercises 1 and 2, which are designed to help you become familiar with and proficient in the use of the compound light microscope. This knowledge will be valuable in later exercises.

Beginning students frequently become impatient with the microscope and forgo this opportunity to practice and develop their observation skills. Simple observation is a critical part of any science. Making discoveries by observation requires *curiosity* and *patience*. We cannot provide procedures for observation, but we can offer this suggestion: Make *careful sketches* to enhance effective observation. You need not be an artist to draw what you see. In your drawings, pay special attention to the following:

1. *Size relationships.* For example, how big are bacteria relative to protozoa?

2. *Spatial relationships.* For example, where is one bacterium in relation to the others? Are they all together in chains?

3. *Behavior.* For example, are individual cells moving, or are they all flowing in the liquid medium?

4. *Sequence of events.* For example, were cells active when you first observed them?

Looking at objects through a microscope is not easy at first, but with a little practice you, like Antoni van Leeuwenhoek, will make discoveries in the microcosms of peppercorn infusions and raindrops (see the photograph below). Van Leeuwenhoek wrote in 1684:

> *Tho my teeth are kept usually very clean, nevertheless when I view them in a Magnifying Glass, I find growing between them a little white matter as thick as wetted flour: In this substance tho I do not perceive any motion, I judged there might probably be living creatures.*
>
> *I therefore took some of this flour and mixed it either with pure rain water wherein were no Animals; or else with some of my Spittle (having no air bubbles to cause a motion in it) and then to my great surprise perceived that the aforesaid matter contained very many small living Animals, which moved themselves very extravagantly. Their motion was strong and nimble, and they darted themselves thro the water or spittle, as a Jack or Pike does thro the water.**

*Quoted in E. B. Fred. "Antoni van Leeuwenhoek." *Journal of Bacteriology* 25:1, 1933.

A replica of the simple microscope made by Antoni van Leeuwenhoek to observe living organisms too small to be seen with the naked eye. The specimen was placed on the tip of the adjustable point and viewed from the other side through the tiny round lens. The highest magnification with his lenses was about 300×.

exercise 1

Use and Care of the Microscope

The most important discoveries of the laws, methods and progress of nature have nearly always sprung from the examination of the smallest objects which she contains.

JEAN BAPTISTE LAMARCK

Objectives

After completing this exercise, you should be able to:

1. Demonstrate the correct use of a compound light microscope.
2. Diagram the path of light through a compound microscope.
3. Name the major parts of a compound microscope.
4. Identify the three basic morphologies of bacteria.

Background

Virtually all organisms studied in microbiology cannot be seen with the naked eye, but require the use of optical systems for magnification. The microscope was invented shortly before 1600 by Zacharias Janssen of the Netherlands. The microscope was not used to examine microorganisms until the 1680s, when a clerk in a dry-goods store, Antoni van Leeuwenhoek, examined scrapings of his teeth and any other substances he could find. The early microscopes, called **simple microscopes,** consisted of biconvex lenses and were essentially magnifying glasses. To see microbes, a compound microscope, which has two lenses between the eye and the object, is required. This optical system magnifies the object, and an illumination system (sun and mirror or lamp) ensures that adequate light is available for viewing. A **brightfield compound microscope,** which shows dark objects in a bright field, is used most often.

You will be using a brightfield compound microscope similar to the one shown in Figure 1.1. The basic frame of the microscope consists of a **base,** a **stage** to hold the slide, an **arm** for carrying the microscope, and a **body tube** for transmitting the magnified image. The stage may have two clips or a movable mechanical stage to hold the slide. The light source is in the base. Above the light source is a **condenser,** which consists of several lenses that concentrate light on the slide by focusing it into a cone, as shown in Figure 1.2. The condenser has an **iris diaphragm,** which controls the angle and

size of the cone of light. This ability to control the *amount* of light ensures that optimal light will reach the slide. Above the stage, on one end of the body tube, is a revolving nosepiece holding three or four **objective lenses.** At the upper end of the tube is an **ocular** or **eyepiece lens** ($10\times$ to $12.5\times$). If a microscope has only one ocular lens, it is called a **monocular** microscope; a **binocular** microscope has two ocular lenses.

By moving the tube closer to the slide or the stage closer to the objective lens, using the coarse or fine adjustment knobs, one can focus the image. The larger knob, the **coarse adjustment,** is used for focusing with the low-power objectives ($4\times$ and $10\times$), and the smaller knob, the **fine adjustment,** is used for focusing with the high-power and oil immersion lenses. The coarse adjustment knob moves the lenses or the stage longer distances. The area seen through a microscope is called the **field of vision.**

The **magnification** of a microscope depends on the type of objective lens used with the ocular. Compound microscopes have three or four objective lenses mounted on a nosepiece: scanning ($4\times$), low-power ($10\times$), high-dry ($40\times$ to $45\times$), and oil immersion ($97\times$ to $100\times$). The magnification provided by each lens is stamped on the barrel. The total magnification of the object is calculated by multiplying the magnification of the ocular (usually $10\times$) by the magnification of the objective lens. The most important lens in microbiology is the **oil immersion lens;** it has the highest magnification ($97\times$ to $100\times$) and must be used with immersion oil. Optical systems could be built to magnify much more than the $1000\times$ magnification of your microscope, but the resolution would be poor.

Resolution or **resolving power** refers to the ability of lenses to reveal fine detail or two points distinctly separated. An example of resolution is a car approaching you at night. At first only one light appears, but as the car nears, you can distinguish two headlights. The resolving power is a function of the wavelength of light used and a characteristic of the lens system called **numerical aperture.** Resolving power is the greatest when two objects are seen as distinct even though they are very close together. Resolving power is expressed in units of length; the smaller the distance, the better the resolving power.

$$\text{Resolving power} = \frac{\text{Wavelength of light used}}{2 \times \text{numerical aperture}}$$

Smaller wavelengths of light increase resolving power. The effect of decreasing the wavelength can be seen in electron microscopes, which use electrons as a source of "light." The electrons have an extremely short wavelength and result in excellent resolving power. A light microscope has a resolving power of about 200 nanometers (nm), whereas an electron microscope has a resolving power of less than 0.2 nm. The numerical aperture is engraved on the side of each objective lens (usually abbreviated N.A.). If the numerical aperture increases—for example, from 0.65 to 1.25—the resolving power is improved. The numerical aperture is dependent on the maximum angle of the light entering the objective lens and on the **refractive index** (the amount the light bends) of the material (usually air) between the objective lens and the slide. This relationship is defined by the following:

N.A. $= N \sin \theta$
$N =$ Refractive index of medium
$\theta =$ Angle between the most divergent light ray gathered by the lens and the center of the lens

As shown in Figure 1.3, light is refracted when it emerges from the slide because of the change in media as the light passes from glass to air. When immersion

Figure 1.1
Photograph of a compound light microscope.

Ocular (eyepiece) A magnifying lens (usually 10X)

Arm

Stage Platform that holds the microscope slide in position

Coarse adjustment knob
Fine adjustment knob
Used for focusing the specimen; turning the knobs changes the distance between the objective lens and the specimen

Base

Body tube Contains mirrors and prisms that transmit the image from the objective lens to the ocular lens

Objective lenses Primary lenses that magnify the specimen

Condenser A lens system that condenses light before it passes through the specimen

Iris diaphragm Controls the amount of light entering the condenser

Mechanical stage knobs Used to move the slide

Light Source of illumination

Figure 1.2
The compound light microscope: its principal parts and their functions. Lines from the light source through the ocular lens illustrate the path of light.

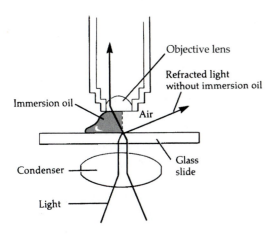

Figure 1.3
Refractive index. Because the refractive indexes of the glass microscope slide and immersion oil are the same, the oil keeps the light rays from refracting.

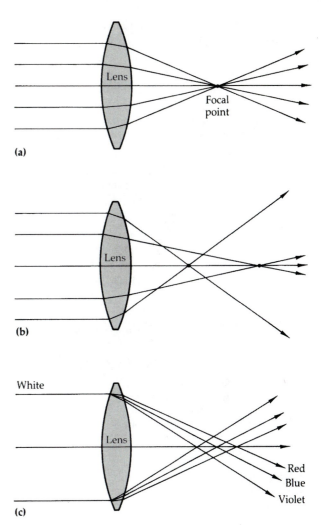

Figure 1.4
Focal point. **(a)** An image is formed when light converges at one point, called the focal point. **(b)** Spherical aberration. Curved lenses result in light passing through one region of the lens having a different focal point than light passing through another part of the lens. **(c)** Chromatic aberration. Each wavelength of light may be given a different focal point by the lens.

oil is placed between the slide and the oil immersion lens, the light ray continues without refraction because immersion oil has the same refractive index ($N = 1.52$) as glass ($N = 1.52$). This can be seen easily. When you look through a bottle of immersion oil, you cannot see the glass rod in it because of the identical N values of the glass and immersion oil. The result of using oil is that light loss is minimized, and the lens focuses very close to the slide.

As light rays pass through a lens, they are bent to converge at the **focal point,** where an image is formed (Figure 1.4a). When you bring the center of a microscope field into focus, the periphery may be fuzzy due to the curvature of the lens, resulting in multiple focal points. This is called **spherical aberration** (Figure 1.4b). Spherical aberrations can be minimized by the use of the iris diaphragm, which eliminates light rays to the periphery of the lens, or by a series of lenses resulting in essentially a flat optical system. Sometimes a multitude of colors, or **chromatic aberration,** is seen in the field (Figure 1.4c). This is due to the prismlike effect of the lens as various wavelengths of white light pass through to a different focal point for each wavelength. Chromatic aberrations can be minimized by the use of filters (usually blue); or by lens systems corrected for red and blue light, called *achromatic lenses;* or by lenses corrected for red, blue, and other wavelengths, called *apochromatic lenses.* The most logical, but most expensive, method of eliminating chromatic aberrations is to use a light source of one wavelength, or **monochromatic light.**

Compound microscopes require a light source. The light may be reflected to the condenser by a mirror under the stage. If your microscope has a mirror, the sun or a lamp may be used as the light source. Most newer compound microscopes have a built-in illuminator in the base. The *intensity* of the light can often be adjusted with a rheostat.

The microscope is a very important tool in microbiology, and it must be used carefully and correctly. Follow these guidelines *every* time you use a microscope.

General Guidelines

1. Carry the microscope with both hands: one hand beneath the base and one hand on the arm.
2. Do not tilt the microscope; instead, adjust your stool so you can comfortably use the instrument.

3. Observe the slide with both eyes open, to avoid eyestrain.
4. Always focus by moving the lens away from the slide.
5. Always focus slowly and carefully.
6. When using the low-power lens, the iris diaphragm should be barely open so that good contrast is achieved. More light is needed with higher magnification.
7. Before using the oil immersion lens, have your slide in focus under high power. *Always focus with low power first.*
8. Keep the stage clean and free of oil. Keep all lenses except the oil immersion lens free of oil.
9. Keep all lenses clean. Use *only* lens paper to clean them. Wipe oil off the oil immersion lens before putting your microscope away. Do not touch the lenses with your hands.
10. Clean the ocular lens carefully with lens paper. If dust is present, it will rotate as you turn the lens.
11. After use, remove the slide, wipe oil off it, put the dust cover on the microscope, and return it to the designated area.
12. When a problem does arise with the microscope, obtain help from the instructor. Do not use another microscope unless yours is declared "out of action."

Materials

Microscope

Immersion oil

Lens paper

Prepared slides of algae, fungi, protozoa, and bacteria

Procedure

1. Place the microscope on the bench squarely in front of you.
2. Obtain a slide of algae or fungi and place it on the stage.
3. Adjust the eyepieces on a binocular microscope to your own personal measurements.
   a. Look through the eyepieces and, using the thumb wheel, adjust the distance between the eyepieces until one circle of light appears.
   b. With the low-power (10×) objective in place, cover the left eyepiece with a small card and focus the microscope on the slide. When the right eyepiece has been focused, remove your hand from the focusing knobs and cover the right eyepiece. Looking through the microscope with your left eye, focus the left eyepiece by turning the eyepiece adjustment. Make a note of the number at which you focused the left eyepiece so you can adjust any binocular microscope for your eyes.
4. Raise the condenser up to the stage. On some microscopes, the condenser can be focused by the following procedure:
   a. Focus with the 10× objective.
   b. Close the iris diaphragm so only a minimum of light enters the objective lens.
   c. Lower the condenser until the light is seen as a circle in the center of the field. On some microscopes the circle of light may be centered (Figure 1.5) using the centering screws found on the condenser.

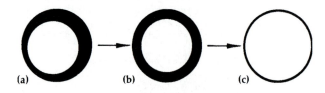

Figure 1.5
Using low power, lower the condenser until a distinct circle of light is visible **(a)**. Center the circle of light using the centering screws **(b)**. Open the iris diaphragm until the light just fills the field **(c)**.

   d. Raise the condenser up to the slide, lower it, and stop when the color on the periphery changes from pink to blue (usually 1 or 2 mm below the stage).
   e. Open the iris diaphragm until the light just fills the field.
5. Diagram some of the cells on the slide under low power. Use a minimum of light by adjusting the _____ .
6. When an image has been brought into focus with low power, rotate the turret to the next lens, and the subject will remain almost in focus. All of the objectives (with the possible exception of the 4×) are **parfocal;** that is, when a subject is in focus with one lens, it will be in focus with all of the lenses. When you have completed your observations under low power, swing the high-dry objective into position and focus. Use the fine adjustment. Only a slight adjustment should be required. Why? _____

More light is usually needed. Again, draw the general size and shape of some cells.

(a) Move the high-dry lens out of position.

(b) Place a drop of immersion oil in the center of the slide.

(c) Move the oil immersion lens into position.

Figure 1.6
Using the oil immersion lens.

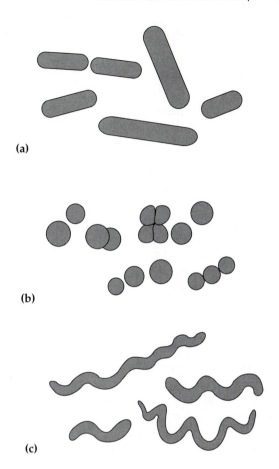

(a)

(b)

(c)

Figure 1.7
Basic shapes of bacteria. **(a)** Bacillus (plural: *bacilli*), or rod. **(b)** Coccus (plural: *cocci*). **(c)** Spirillum (plural: *spirilla*).

7. Move the high-dry lens out of position, and place a drop of immersion oil on the area of the slide you are observing. Carefully click the oil immersion lens into position. It should now be immersed in the oil (Figure 1.6). Careful use of the fine adjust- ment knob should bring the object into focus. Note the shape and size of the cells. Did the color of the cells change with the different lenses? _____ Did the size of the field change? _____

8. The following figures may be used as references:
Algae: Plate XIII
Protozoa: Figure 37.1
Fungi: Figure 35.1

9. When your observations are completed, move the turret to bring a low-power objective into posi- tion. *Do not* rotate the high-dry (40×) objective through the immersion oil. Clean the oil off the objective lens with lens paper, and clean off the slide with tissue paper or a paper towel. Remove the slide. Repeat this procedure with all the avail- able slides. When observing the bacteria, note the three different morphologies, or shapes, shown in Figure 1.7.

exercise 1

Use and Care of the Microscope

Name _____

Date _____

Lab Section _____

Purpose _____

_____

_____

Data

Microscope number: _____ Monocular or binocular: _____

Eyepiece adjustment notes: _____

Draw a few representative cells from each slide and show how they appeared at each magnification. Note the differences in size at each magnification.

Algae

Slide of _____

Total magnification: <u>10</u> ×　　　　　　<u>40</u>×　　　　　　<u>100</u> ×

Amoeba proteus

10x　　　　　　40x　　　　　　100x

Fungi

Slide of _____

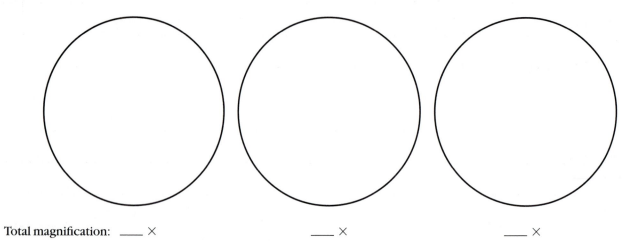

Total magnification: _____ ×   _____ ×   _____ ×

Protozoa

Slide of _____

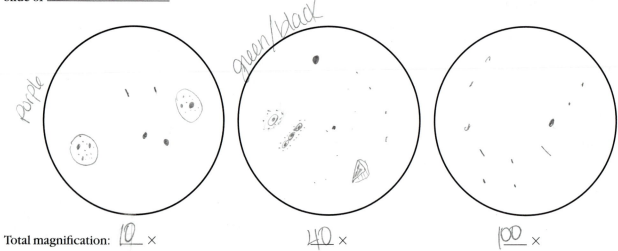

Total magnification: 10 ×   40 ×   100 ×

Bacteria

Slide of _____

Be sure to sketch all bacterial shapes observed.

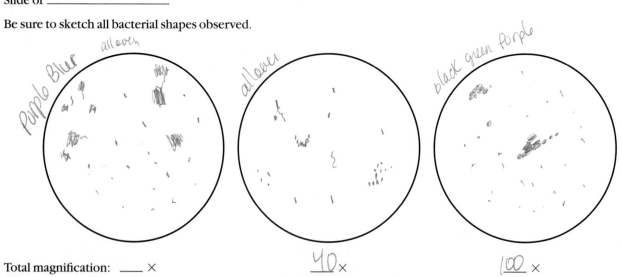

Total magnification: _____ ×   40 ×   100 ×

Questions

1. What does it mean when a microscope is parfocal? _It means the subject being observed under one power when focused shouldn't need much adjustment when switching to another power. or lenses_

2. Which objective focuses closest to the slide? _____

3. What controls the amount of light reaching the ocular lens? _the condenser and iris diaphragm condenser has several lenses that concentrate light on the slide by focusing it into a cone. Iris diaphragm controls the angle size of the cone of light_

4. Is your lens corrected for chromatic aberrations? _____

5. What three bacterial shapes did you observe? _____

Critical Thinking

1. Name two ways in which you can enhance the resolving power.

2. What advantages does the low-power objective have over the oil immersion objective for fungi?

3. What would occur if water were accidentally used in place of immersion oil?

exercise 2

Examination of Living Microorganisms

Objectives

After completing this exercise, you should be able to:

1. Prepare and observe wet mount slides and hanging-drop slides.
2. Distinguish between true motility and Brownian movement.
3. Use a phase-contrast microscope.
4. Explain how phase-contrast and darkfield microscopy differ from brightfield microscopy.

Background

Antoni van Leeuwenhoek was the first known individual to observe living microbes in a suspension. Unfortunately, he was very protective of his homemade microscopes and left no descriptions of how to make them. During his lifetime he kept "for himself alone" his microscopes and his method of observing "animalcules." Directions for making a replica of van Leeuwenhoek's microscope can be found in *American Biology Teacher.**

In **brightfield microscopy,** objects are dark and the field is light. Brightfield microscopy can be used to observe unstained microorganisms. However, since the optical properties of the organisms and their aqueous environment are similar, very little contrast can be seen. Two other types of compound microscopes, however, are useful for observing living organisms: darkfield and phase-contrast microscopes. These microscopes optically increase contrasts between the organism and background by using special condensers.

In **darkfield microscopy,** the objects are light and the field is dark. In brightfield microscopy, light rays that strike the specimen are reflected away from the lens (Figure 2.1a). The darkfield condenser concentrates the light into a hollow cone of light at such an angle that none of the light rays reach the objective lens unless they pass through an object such as a cell to change their direction (Figure 2.1b). An opaque darkfield disk eliminates all central light rays. Thus, the objects appear brightly illuminated against a dark background. Darkfield microscopy allows the investigator to observe the shape and motility of unstained live or-

ganisms. Darkfield microscopy is valuable for observing the spirochete (*Treponema pallidum*) that causes syphillis. This bacterium is not stainable with conventional stains but can be observed in direct smears with darkfield microscopy.

In **phase-contrast microscopy,** small differences in the refractive properties of the objects and the aqueous environment are transformed into corresponding variations of brightness. In phase-contrast microscopy, a ring of light passes through the object, and light rays are **diffracted** (retarded) and out of phase with the light rays not hitting the object. The phase-contrast microscope enhances these phase differences so that the eye detects the difference as contrast (Figure 2.2) between the organisms and background and between structures within a cell. In phase-contrast microscopy, the organisms appear as degrees of brightness against a darker background. The advantage of phase-contrast microscopy is that structural detail within live cells can be studied.

In this exercise, you will examine, using wet mount techniques, different environments to help you become aware of the numbers and varieties of microbes found in nature. The microbes will exhibit either Brownian movement or true motility. **Brownian movement** is not true motility but rather movement caused by the molecules in the liquid striking an object and causing the object to shake or bounce. In Brownian movement, the particles and microorganisms all vibrate at about the same rate and maintain their relative positions. Motile microorganisms move from one position to another. Their movement appears more directed than Brownian movement, and occasionally the cells may spin or roll.

Many kinds of microbes, such as protozoa, algae, fungi, and bacteria, can be found in pond water and in infusions of organic matter. Van Leeuwenhoek made some of his discoveries using a peppercorn infusion similar to the one you will see in this exercise. Direct examination of living microorganisms is very useful in determining size, shape, and movement. A wet mount is a fast way to observe bacteria. Motility and larger microbes are more easily observed in the greater depth provided by a hanging drop. Evaporation of the suspended drop of fluid is reduced by using a petroleum jelly seal.

In this exercise, you will examine living microorganisms by brightfield and by phase-contrast microscopy.

*W. G. Walter and H. Via. "Making a Leeuwenhoek Microscope Replica." *American Biology Teacher* 30 (6):537–539, 1968.

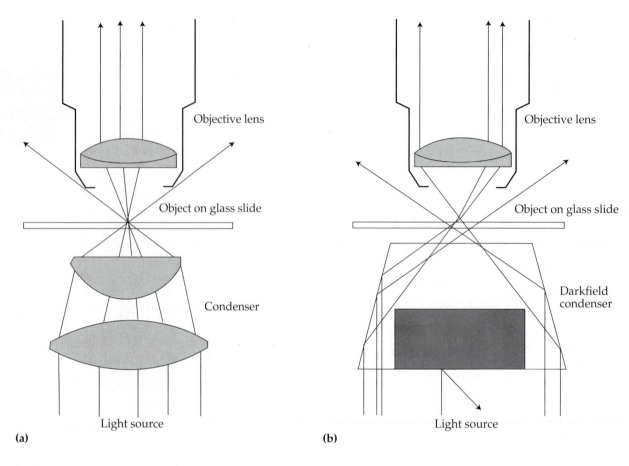

(a)　　　　　　　　　　　　　　**(b)**

Figure 2.1

A comparison of brightfield and darkfield microscopy: **(a)** In brightfield microscopy, light is reflected away from the objective lens by the specimen. **(b)** In darkfield microscopy, only light rays that go through the object reach the objective lens.

Materials

Slides

Coverslips

Hanging-drop (depression) slide

Petroleum jelly

Toothpick

Pasteur pipettes

Alcohol

Gram's iodine

Phase-contrast microscope and centering telescope

Cultures

Hay infusion, incubated 1 week in light

Hay infusion, incubated 1 week in dark

Peppercorn infusion

Pond water with algae

12- to 18-hour-old broth culture of *Bacillus*

Techniques Required

Compound light microscopy, Exercise 1

Procedure

Wet Mount Technique

1. Suspend the infusions by stirring or shaking them carefully. Using a Pasteur pipette, transfer a small drop of one hay infusion to a slide.
2. Handle the coverslip carefully by its edges and place it on the drop.
3. Gently press on the coverslip with the end of a pencil.
4. Place the slide on the microscope stage and observe it with low power (10×). Adjust the iris diaphragm so a small amount of light is admitted. Concentrate your observations on the larger, more rapidly moving organisms. At this magnification, bacteria are barely discernible as tiny dots. Figure 2.3 may help you to identify some of the microorganisms.
5. Examine the slide with the high-dry lens (40×), then increase the light and focus carefully. Bac-

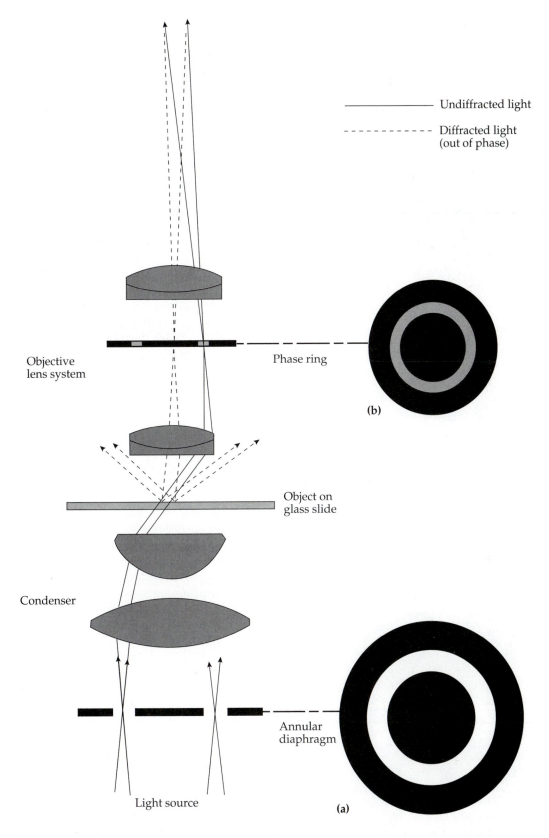

Figure 2.2
Phase-contrast microscope. A hollow cone of light is formed by the annular diaphragm **(a)**. Diffracted rays are further retarded by the phase ring **(b)** in the objective lens.

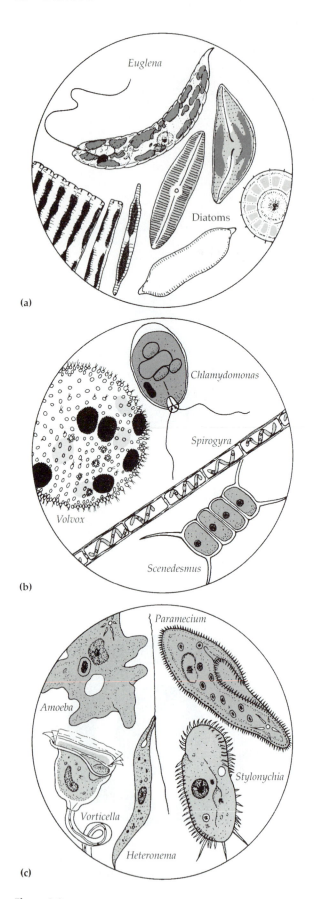

(a)

(b)

(c)

Figure 2.3
Some common protozoa and algae that can be found in infusions. **(a)** and **(b)** Algae. **(c)** Protozoa.

teria should now be magnified sufficiently to be seen.

6. After recording your observations, examine the slide with the oil immersion lens. Some microorganisms are motile, while others exhibit Brownian movement.

7. If you want to observe the motile organisms further, place a drop of alcohol or Gram's iodine at the edge of the coverslip and allow it to run under and mix with the infusion. What does the alcohol or iodine do to these organisms? _____

_____

They can now be observed more carefully.

8. Record your observations, noting the relative size and shape of the organisms.

9. Make a wet mount from the other hay infusion, and observe it, using the low and high-dry objectives. Record your observations.

10. Clean all the slides and return them to the slide box. Coverslips can be discarded in the disinfectant jar. Wipe the oil from the objective lens with lens paper.

Hanging-Drop Procedure

1. Obtain a depression slide.

2. Pick up a small amount of petroleum jelly on a toothpick.

3. Pick up a coverslip (by its edges) and carefully scrape the petroleum jelly with an edge of the coverslip to get a small rim of petroleum jelly. Repeat with the other three edges (Figure 2.4a), keeping the petroleum jelly on the same side of the coverslip.

4. Place the coverslip on a paper towel, with the petroleum jelly–side up.

5. Transfer a small drop or loopful of the peppercorn infusion to the coverslip.

6. Place a slide over the drop and quickly invert it so the drop is suspended (Figure 2.4b and c). Why should the drop be hanging? _____

_____

7. Examine the drop under low power (Figure 2.4d) by locating the edge of the drop and moving the slide so the edge of the drop crosses the center of the field.

8. Reduce the light with the iris diaphragm and focus. Observe the different sizes, shapes, and types of movement.

9. Switch to high-dry and record your observations. Do not focus down. Why not? _____

_____

Do not attempt to use the oil immersion lens.

10. When finished, clean your slide and, using a new coverslip, repeat the procedure with the culture of *Bacillus*. Record your observations.

11. Return your microscope to its proper location. Clean your slides well and return them. Discard the coverslips in the disinfectant jar.

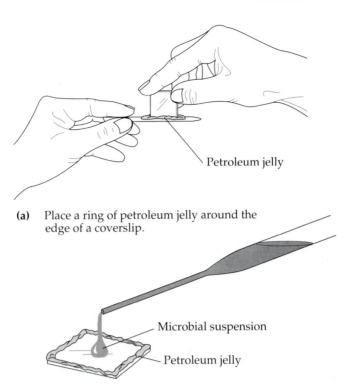

(a) Place a ring of petroleum jelly around the edge of a coverslip.

(b) Place a drop of an infusion in the center of the coverslip.

(c) Place the depression slide on the coverslip.

(d) Turn the slide over, place the slide, coverslip up, on the microscope stage, and observe it under the low and high-dry objectives.

Figure 2.4
Hanging-drop preparation.

Figure 2.5

Adjustment of phase rings. (**a**) The two images seen before adjustment of the phase rings. By moving the wheel under the condenser turret (**b**), the images are made to coincide (**c**).

Phase-Contrast Microscopy

1. Make a wet mount of any of the suspensions.
2. Place the slide on the stage and turn on the light.
3. Start with the 10× objective and move the condenser diaphragm to match setting "10."
4. Focus on an obvious clump of material with the coarse and fine adjustments.
5. Close the iris diaphragm, and move the condenser up and down until a light octagon comes into focus. Then open the diaphragm until light just fills the field.
6. Adjust the phase rings.
   a. Replace the eyepiece (screw out) with the centering telescope, and focus the telescope on the phase plate ring by revolving its head.
   b. You will see two rings (Figure 2.5a): one, a bright image of the phase ring, and the other, the dark image of the phase plate in the objective. Adjust the knurled wheel under the condenser turret with your fingertips (Figure 2.5b) to make the bright image coincide with the dark ring (Figure 2.5c). Do not touch the diaphragm.
   c. When the ring has been centered (Figure 2.5c), replace the telescope with the ocular lens.
7. Observe.
8. Focus the slide with the 40× objective and the 40× condenser turret.
9. Readjust the phase rings using the telescope, as done in step 6. Record your observations. Can you distinguish any of the organelles in the organisms? _____
10. Focus the slide with the 100× objective and the 100× condenser turret, then readjust the phase rings using the telescope. Diagram your observations.
11. Clean the slide. Make a wet mount of one other sample. Observe. Is it motile? _____
    Compare your observations with those made with the brightfield microscope.

exercise 2

Examination of Living Microorganisms

Name _____

Date _____

Lab Section _____

Purpose _____

_____

_____

Data

Wet Mount Technique

Draw the types of protozoa, algae, fungi, and bacteria observed. Indicate their relative sizes and shapes. Record the magnification.

Hay infusion, light

Hay infusion, dark

Total magnification: 40×   40×

Hanging-Drop Procedure

Draw the types of microorganisms seen under high-dry magnification. Indicate their relative sizes and shapes.

Peppercorn infusion

Bacillus culture

Pond water

Total magnification: 40× ___ × ___ ×

Phase-Contrast Microscopy

Carefully draw the organisms and their internal structure.

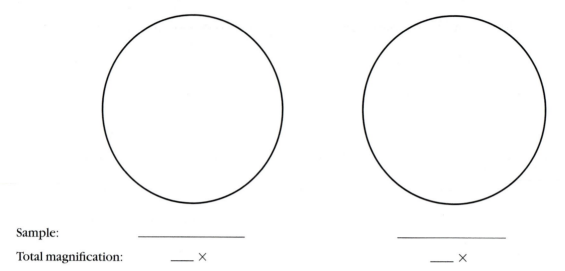

Sample: _____   _____

Total magnification: ___ ×   ___ ×

Bacteria

In the table below, record the relative numbers of each bacterial shape observed. Record your data as 4+ (most abundant), 3+, 2+, +, − (not seen).

Culture	Shape		
	Bacilli (rods)	Cocci	Spirilla
Hay infusion, dark			
Hay infusion, light			
Peppercorn infusion			
Bacillus culture			
Pond water			

Questions

1. What, if any, practical value do these techniques have? _____

_____

2. Did any of the bacteria exhibit true motility? _____ How do you distinguish true motility from Brownian

movement or motion of the fluid? _____

_____

3.   What is the advantage of the hanging-drop procedure over the wet mount? _____

_____

4.   Compare the appearance of microorganisms observed using phase-contrast microscopy versus brightfield

microscopy. _____

_____

_____

5.   Why is petroleum jelly used in the hanging-drop procedure? _____

_____

_____

6.   Which infusion had the most bacteria? _____ Briefly explain why. _____

_____

_____

_____

Critical Thinking

1.   Why are microorganisms hard to see in wet preparations?

2.   Can you distinguish the prokaryotic organisms from the eukaryotic organisms? Explain.

3.   Why isn't the oil immersion lens used in the hanging-drop procedure?

4.   From where did the organisms in the infusions come?

Staining Methods

EXERCISES

In 1877, Robert Koch wrote:

> *How many incomplete and false observations might have remained unpublished instead of swelling the bacterial literature into a turbid stream, if investigators had checked their preparations with each other?*[*]

To solve this problem, he introduced into microbiology the procedures of air drying, chemical fixation, and staining with aniline dyes.

In addition to making a lasting preparation, staining bacteria enhances the contrast between bacteria and the surrounding material and permits observation of greater detail and resolution than wet mount procedures. Microorganisms are prepared for staining by smearing them onto a microscope slide.

In these exercises, bacteria will be transferred from growth or culture media to microscope slides using an inoculating loop. An **inoculating loop** is a Nichrome wire held with an insulated handle. Before and after it is used,

*Quoted in H. A. Lechevalier and M. Solotorovsky. *Three Centuries of Microbiology.* New York: Dover Publications, 1974, p. 79.

the inoculating loop is sterilized by heating or **flaming**—that is, the loop is held in the flame of a burner (photograph a) or electric incinerator (photograph b) until it is red-hot.

Staining techniques may involve **simple stains,** in which only one reagent is used and all bacteria are usually stained similarly (Exercises 3 and 4), or **differential stains,** in which multiple reagents are used and bacteria react to the reagents differently (Exercises 5 and 6). Structural stains are used to identify specific parts of microorganisms (Exercise 7).

Most of the bacteria that are grown in laboratories are rods or cocci (see color plates I.1 and I.2). Rods and cocci will be used throughout this part. Staining and microscopic examination are usually the first steps in identifying microorganisms. In Exercise 8, Morphologic Unknown, you will be asked to characterize selected bacteria by their microscopic appearance.

Sterilizing an inoculating loop. **(a)** Before and after use, sterilize the loop with a flame until it is red-hot. **(b)** The loop can be sterilized in an electric incinerator.

(b)

(a)

exercise 3

Preparation of Smears and Simple Staining

Objectives

After completing this exercise, you should be able to:

1. Make and fix a smear.
2. List the advantages of staining microorganisms.
3. Explain the basic mechanism of staining.
4. Perform a simple direct stain.

Background

Most stains used in microbiology are synthetic aniline (coal tar derivative) dyes derived from benzene. The dyes are usually salts, although a few are acids or bases, composed of charged colored ions. The ion that is colored is referred to as a **chromophore.** For example,

$$\text{Methylene blue chloride} \leftrightarrows \text{Methylene blue}^+ + \text{Cl}^-$$
$$\text{(Chromophore)}$$

If the chromophore is a positive ion like the methylene blue in the equation shown, the stain is considered a **basic stain;** if a negative ion, it is an **acidic stain.** Most bacteria are stained when a basic stain permeates the cell wall and adheres by weak ionic bonds to the negative charges of the bacterial cell.

Staining procedures that use only one stain are called **simple stains.** A simple stain that stains the bacteria is a **direct stain,** and a simple stain that stains the background but leaves the bacteria unstained is a **negative stain.** Simple stains can be used to determine cell morphology, size, and arrangement.

Before bacteria can be stained, a smear must be made and fixed. A **smear** is made by spreading a bacterial suspension on a clean slide and allowing it to air dry. The dry smear is passed through a Bunsen burner flame several times to **heat fix** the bacteria. Heat fixing in a flame may not kill all the bacteria. Alternatively, the dry smear can be placed on a 60°C slide warmer for 10 minutes or chemically fixed. To **chemically fix** the bacteria, cover the smear with 95% methyl alcohol for 1 minute. Fixing denatures bacterial enzymes, preventing them from digesting cell parts, which causes the cell to break, a process called *autolysis.* Fixing also enhances the adherence of bacterial cells to the microscope slide.

Materials

Methylene blue

Wash bottle of distilled water

Slide

Inoculating loop

Cultures

Staphylococcus epidermidis slant

Bacillus megaterium broth

Techniques Required

Compound light microscopy, Exercise 1

Inoculating loop, pp. 27–28

Procedure

1. Clean your slide well with abrasive soap or cleanser; rinse and dry. Handle clean slides by the end or edge. Use a marker to make two dime-sized circles on each slide, on the bottom of the slide so they will not wash off. Label each circle according to the bacterial culture used.

2. For the bacterial culture on solid media, place 1 or 2 loopfuls of distilled water in the center of one circle, using the inoculating loop. Which bacterium is on a solid medium? _____

3. Sterilize your inoculating loop by holding it in the hottest part of the flame (at the edge of the inner blue area) until it is red-hot (see the figure on the facing page). The entire wire should get red. Allow the loop to cool so that bacteria picked up with the loop won't be killed. Allow the loop to cool without touching it. Cooling takes about 30 seconds. You will determine the appropriate time with a little practice.

 The loop must be cool before inserting it into a medium. A loop that is too hot will spatter the medium and move bacteria into the air.

4. Smear preparation (Figure 3.1).
   a. Using the cooled loop, scrape a *small* amount of the culture off the slant. If you hear the sizzle of boiling water when you touch the agar with

(a) Mark the smear areas with a marking pencil on the underside of a clean slide.

FROM SOLID MEDIUM

FROM LIQUID MEDIUM

(b) Place 1 or 2 loopfuls of water on the slide.

(d) Place 2 or 3 loopfuls of the liquid culture on the slide with a sterile loop.

(c) Transfer a very small amount of the culture with a sterile loop. Mix with the water on the slide.

(e) Spread the bacteria within the circle.

(f) Allow the smears to air dry at room temperature.

or

(g) Pass the slide through the flame of a burner two or three times.

(h) Cover the smears with 95% methyl alcohol for 1 minute, then let the smears air dry.

Figure 3.1

Preparing a bacterial smear.

the loop, reflame your loop and begin again. Why? _____

Try not to gouge the agar. Emulsify (to a milky suspension) the cells in the drop of water, and spread the suspension to fill a majority of the circle. The smear should look like diluted skim milk. Flame your loop again.

b. Make a smear of bacteria from the broth culture on the other circle. *Do not use water,* because the bacteria are already suspended in water. Flick the tube of broth culture lightly with your finger to resuspend sedimented bacteria, and place 2 or 3 loopfuls of the culture in the circle. Flame your loop between each loopful. Spread the culture within the circle (Figure 3.2).

c. Flame your loop.

🔬 **Always flame your loop after using it and before setting it down.**

---

d. Let the smears dry. *Do not* blow on the slide, as this will move the bacterial suspension. *Do not* flame the slide, as flaming will distort the cells' shapes.

e. Hold the slide with a clothespin and fix the smears by one of the following methods (Figure 3.1g or h):

(1) Pass the slide quickly through the blue flame two or three times.

(2) Cover the smear with 95% methyl alcohol for 1 minute. Tip the slide to let the alcohol run off, and let the slide air dry before staining. Do not fix until the smears are completely dry. Why? _____

_____

5. Staining (Figure 3.3).

a. Use a clothespin to hold the slide, or place it on a staining rack.

b. Cover the smear with methylene blue and leave it for 30 to 60 seconds (Figure 3.3a).

c. Carefully wash the excess stain off with distilled water from a wash bottle. Let the water run down the tilted slide (Figure 3.3b).

Figure 3.2

A loopful of microbial suspension is transferred to a slide.

d. Gently blot the smear with a paper towel or absorbent paper and let it dry (Figure 3.3c and d).

6. Examine your stained smears microscopically using the low, high-dry, and oil immersion objectives. Put the oil *directly* on the smear; coverslips are not needed. Record your observations with labeled drawings.

7. Blot the oil from the objective lens with lens paper, and return your microscope to its proper location. Clean your slides well, or save them as described in step 8.

8. Stained bacterial slides can be stored in a slide box. Remove the oil from the slide by blotting it with a paper towel. Any residual oil won't matter.

(a) Cover the smear with methylene blue for 30 seconds.

(c) Wash the smear with water.

(b) Gently wash off the methylene blue with water by squirting the water so it runs through the smear.

(d) Blot it dry.

Figure 3.3
Simple staining.

exercise 3

Preparation of Smears and Simple Staining

Name _____

Date _____

Lab Section _____

Purpose _____

_____

_____

Data

Staphylococcus epidermidis

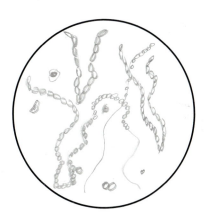

Sketch a few bacteria under
1000× magnification.

Morphology (shape): _____

Arrangement of cells
relative to one another: _____

Bacillus megaterium

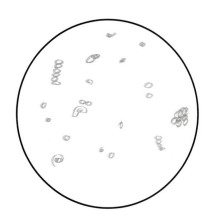

Sketch a few bacteria under
1000× magnification.

Morphology (shape): _____

Arrangement of cells
relative to one another: _____

Questions

1. Which bacterium is a rod? _____

2. Of what value is a simple stain? _____

   _____

3. What is the purpose of fixing the smear? _____

   _____

4. In heat fixing, what would happen if too much heat were applied? _____

   _____

Critical Thinking

1. Methylene blue can be prepared as a basic stain or an acidic stain. How would the pH affect the staining of bacteria?

2. Can dyes other than methylene blue be used for direct staining? Briefly explain.

3. Bacteria can be seen without staining. Why then was Koch's recommendation for fixing and staining important for microbiology?

exercise 4

Negative Staining

Objectives

After completing this exercise, you should be able to:

1. Explain the application and mechanism of the negative stain technique.
2. Prepare a negative stain.

Background

The **negative stain** technique does not stain the bacteria but stains the background. The bacteria will appear clear against a stained background. The stain does not stain the bacteria because of ionic repulsion: The bacteria and the acidic stain both have negative charges.

No heat fixing or strong chemicals are used, so the bacteria are less distorted than in other staining procedures. The negative stain technique is very useful in situations where other staining techniques don't clearly indicate cell morphology or size.

Materials

Nigrosin

Clean slides (6)

Distilled water

Sterile toothpicks

Cultures

Bacillus subtilis

Staphylococcus epidermidis

Techniques Required

Compound light microscopy, Exercise 1

Smear preparation, Exercise 3

Procedure (Figure 4.1)

1. Slides must be clean and grease-free.
2. **a.** Place a *small* drop of nigrosin at the end of the slide. For cultures on solid media, add a loopful of distilled water and emulsify a small amount of the culture in the nigrosin-water drop. For broth cultures, mix a loopful of the culture into the drop of nigrosin (Figure 4.1a). Do not spread the drop or let it dry.
   b. Using the end edge of another slide, spread the drop out (Figure 4.1b and c) to produce a smear varying from opaque black to gray. The angle of the spreading slide will determine the thickness of the smear.
   c. Let the smear air dry (Figure 4.1d). *Do not heat-fix it.*
   d. Prepare a negative stain of the other culture.
3. Examine the stained slides microscopically using the low, high-dry, and oil immersion objectives (Figure 4.2). As a general rule, if a few short rods are seen with small cocci, the morphology is rod-shaped. The apparent cocci are short rods viewed from the end or products of the cell division of small rods. (See color plate I.4.)
4. **a.** Place a small drop of nigrosin and a loopful of water at the end of a slide.
   b. Scrape the base of your teeth and gums with a sterile toothpick.
   c. Mix the nigrosin-water with the toothpick to get an emulsion of bacteria from your mouth.

> **☣ Discard the toothpick in disinfectant.**

   d. Follow steps 2b and 2c above to complete the negative stain. Observe the stain and describe your results.
5. Wash your slides. Wipe the oil off your microscope and return it.

(a)   Place a small drop of nigrosin near one end of a slide. Mix a loopful of broth culture in the drop. When the organisms are taken from a solid medium, mix a loopful of water in the nigrosin.

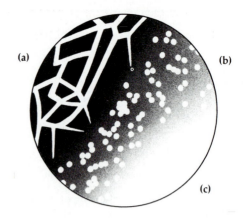

Figure 4.2
A negative stain viewed under a microscope. **(a)** This part of the smear is too heavy: Cracks in the stain can be seen. **(b)** Colorless cells are visible here. **(c)** Too little stain is in this area of the smear.

(b)   Gently draw a second slide across the surface of the first until it contacts the drop. The drop will spread across the edge of the top slide.

(c)   Push the top slide to the left along the entire surface of the bottom slide.

(d)   Let the smear air dry.

Figure 4.1
Preparing a negative stain.

exercise 4

Negative Staining

Name _____

Date _____

Lab Section _____

Purpose _____
_____
_____

Data

Sketch a few bacteria (1000×).

Bacillus subtilis

Morphology and
arrangement: _____

Staphylococcus epidermidis

Morphology and
arrangement: _____

Tooth and gum
scraping

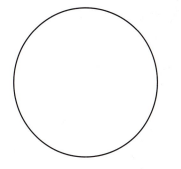

Morphology and
arrangement: _____

Questions

1. Which cell is a rod? _____ How does its appearance differ from the rod you

    stained with methylene blue (simple stain)? _____

2. Why is the size more accurate in a negative stain than in a simple stain? _____

    _____

3. Could any dye be used in place of nigrosin for negative staining? _____. What types of dyes are used for

    negative staining? _____

    Briefly explain. _____

    _____

Critical Thinking

1. What microscopic technique gives a field similar in appearance to that seen in the negative stain?

2. Carbolfuchsin can be used as a simple stain and as a negative stain. As a simple stain, the pH is _____.

3. India ink can give the appearance of a negative stain when used in a wet mount. What is the basis (e.g., pH) for
    this stain? (India ink is a suspension of 0.5–1.0 µm carbon particles in water.)

exercise 5

Gram Staining

Objectives

After completing this exercise, you should be able to:

1. Explain the rationale and procedure for the Gram stain.
2. Perform and interpret Gram stains.

Background

The Gram stain is a very useful stain for identifying and classifying bacteria. The **Gram stain** is a differential stain that allows you to classify bacteria as either gram-positive or gram-negative. The Gram staining technique was discovered by Hans Christian Gram in 1884, when he attempted to stain cells and found that some lost their color when excess stain was washed off.

The staining technique consists of the following steps:

1. Apply **primary stain** (crystal violet). All bacteria are stained purple by this basic dye.
2. Apply **mordant** (Gram's iodine). The iodine combines with the crystal violet in the cell to form a crystal violet–iodine complex (CV–I).
3. Apply **decolorizing agent** (ethyl alcohol or ethyl alcohol–acetone). The primary stain is washed out (decolorized) of some bacteria, while others are unaffected.
4. Apply **secondary stain** or **counterstain** (safranin). This basic dye stains the decolorized bacteria red.

Many variations of the Gram stain have been developed. The most important determining factor in the procedure is that bacteria differ in their *rate* of decolorization. Those that decolorize easily are referred to as **gram-negative,** whereas those that retain the primary stain are called **gram-positive.**

Bacteria stain differently because of chemical and physical differences in their cell walls. Crystal violet is picked up by the cell. Iodine reacts with the dye in the cytoplasm to form CV–I. In gram-negative cells, the decolorizing agent dissolves the outer lipopolysaccharide layer, and the CV–I washes out through the thin layer of peptidoglycan.

The Gram stain is most consistent when done on young cultures of bacteria (less than 24 hours old). Because Gram staining is usually the first step in iden-ᶠʸing bacteria, the procedure should be memorized.

Materials

Gram staining reagents:
　　Crystal violet
　　Gram's iodine
　　Ethyl alcohol
　　Safranin

Wash bottle of distilled water

Slide

Cultures

Staphylococcus epidermidis cocci positive

Escherichia coli rod negative

Bacillus subtilis rod positive

Techniques Required

Compound light microscopy, Exercise 1

Smear preparation, Exercise 3

Simple staining, Exercise 3

Procedure (Figure 5.1)

1. Prepare and fix smears. Clean the slide well, and make three circles on it with a marker. Label each circle for one of the cultures.
2. Prepare a Gram stain. Use a clothespin or slide rack to hold the slide.
   a. Cover the smear with crystal violet and leave it for 30 seconds (Figure 5.1a).
   b. Wash the slide carefully with distilled water from a wash bottle. Do not squirt water directly onto the smear (Figure 5.1b).
   c. Cover the smear with Gram's iodine for 10 seconds (Figure 5.1c).
   d. Wash off the iodine by tilting the slide and squirting water above the smear so that the water runs over the smear (Figure 5.1d).
   e. Decolorize with 95% ethyl alcohol (Figure 5.1e). Let the alcohol run through the smear until no large amounts of purple wash out (usually 10 to 20 seconds). The degree of alcohol decolorizing depends on the thickness of the smear. This is a critical step. *Do not over-decolorize.* However, experience is the only

(a)   Cover the smear with crystal violet for 30 seconds.

(b)   Gently wash off the crystal violet with water by squirting the water so it runs through the smear.

(c)   Cover the smear with Gram's iodine for 10 seconds.

(d)   Gently wash the smear with water.

(e)   Decolorize with alcohol.

(f)   Gently wash off the alcohol.

(g)   Cover the smear with safranin for 30 seconds.

(h)   Wash the smear with water.

(i)   Blot it dry.

Figure 5.1
The Gram stain.

way you will be able to determine how long to decolorize for.

f. Immediately wash gently with distilled water (Figure 5.1f). Why? _____

g. Add safranin for 30 seconds (Figure 5.1g).

h. Wash the slide with distilled water and blot it dry with a paper towel or absorbent paper (Figure 5.1h and i).

3. Examine the stained slide microscopically using the low, high-dry, and oil immersion objectives. Put the oil directly on the smear. Record your observa-tions. (See color plates I.1 and I.2.) Do they agree with those given in your textbook? _____ If not, try to determine why. Some common sources of Gram staining errors are the following:

a. The loop was too hot.

b. Excessive heat was applied during heat fixing.

c. The decolorizing agent (alcohol) was left on the smear too long.

4. Design your own differential stain using different reagents but the same general scheme. Does it work? _____

exercise 5

Gram Staining

Name _____

Date _____

Lab Section _____

Purpose _____
_____
_____

Data

	Staphylococcus epidermidis	*Bacillus subtilis*	*Escherichia coli*

negative rods

Sketch a few bacteria (1000×).

Morphology and arrangement: _____ _____ _____

Color: _____ _____ _____

Gram reaction: _____ _____ _____

Which organism is the largest? _____ The smallest? _____

Questions

1. Did your results agree with the information in your textbook? _____ If not, why not?
_____
_____

2. Why will old gram-positive cells stain gram-negative? _? weak cell wall allowing crystal violet
dye to leak out_
_____

3.   Can iodine be added before the primary stain in a Gram stain? _____

_____

_____

4.   List the steps of the Gram staining procedure in order (omit washings), and fill in the color of gram-positive cells and gram-negative cells after each step.

Step	Chemical	Appearance	
		Gram-Positive Cells	Gram-Negative Cells
1			
2			
3			
4			

5.   Which step can be omitted without affecting determination of the Gram reaction? _____

Critical Thinking

1.   Suppose you Gram stained a sample from a pure culture of bacteria and observed a field of red and purple cocci. Adjacent cells were not always the same color. What do you conclude?

2.   Suppose you are viewing a Gram stained field of red rods and purple cocci through the microscope. What do you conclude?

3.   Since you can't identify bacteria from a Gram stain, why might a physician perform a Gram stain on a sample before prescribing an antibiotic?

4.   If you Gram stained human cells, what would happen?

exercise 6

Acid-Fast Staining

Objectives

After completing this exercise, you should be able to:

1. Apply the acid-fast procedure.
2. Explain what is occurring during the acid-fast staining procedure.
3. Perform and interpret an acid-fast stain.

Background

The **acid-fast stain** is a differential stain. In 1882 Paul Ehrlich discovered that *Mycobacterium tuberculosis* (the causative agent of tuberculosis) retained the primary stain even after washing with an acid-alcohol mixture. (We hope you can appreciate the phenomenal strides that were made in microbiology in the 1880s. Most of the staining and culturing techniques used today originated during that time.) Most bacteria are decolorized by acid-alcohol, with only the families Mycobacteriaceae, Nocardiaceae, Goroniaceae, and Tsukamurellaceae (*Bergey's Manual**) being acid-fast. The acid-fast technique has great value as a diagnostic procedure because both *Mycobacterium* and *Nocardia* contain **pathogenic** (disease-causing) species.

The cell walls of acid-fast organisms contain a wax-like lipid called **mycolic acid,** which renders the cell wall impermeable to most stains. The cell wall is so impermeable, in fact, that a clinical specimen is usually treated with strong sodium hydroxide to remove debris and contaminating bacteria prior to culturing mycobacteria. The mycobacteria are not killed by this procedure.

Today, the techniques developed by Franz Ziehl and Friedrich Neelsen and by Joseph J. Kinyoun are the most widely used acid-fast stains. In the **Ziehl–Neelsen procedure,** the smear is flooded with carbolfuchsin (a dark red dye containing 5% phenol), which has a high affinity for a chemical component of the bacterial cell. The smear is heated to facilitate penetration of the stain into the bacteria. The stained smears are washed with an acid-alcohol mixture that easily decolorizes most bacteria except the acid-fast microbes. Methylene blue

is then used as a counterstain to enable you to observe the non-acid-fast organisms. In the **Kinyoun modification**, called a *cold stain*, the concentrations of phenol and carbolfuchsin are increased so heating isn't necessary.

The mechanism of the acid-fast stain is probably the result of the relative solubility of carbolfuchsin and the impermeability of the cell wall. Fuchsin is more soluble in carbolic acid (phenol) than in water, and carbolic acid solubilizes more easily in lipids than in acid-alcohol. Therefore, carbolfuchsin has a higher affinity for lipids than for acid-alcohol and will remain with the cell wall when washed with acid-alcohol.

Materials

Acid-fast staining reagents:
    Kinyoun's carbolfuchsin
    Acid-alcohol
    Methylene blue

Wash bottle of distilled water

Slide

Cultures

Mycobacterium phlei

Escherichia coli

Demonstration Slides

Acid-fast sputum slides

Techniques Required

Compound light microscopy, Exercise 1

Smear preparation, Exercise 3

Simple staining, Exercise 3

Procedure

1. Prepare and fix a smear of each culture.
2. Cover the smears with carbolfuchsin and leave them for 5 minutes (Figure 6.1a).
3. Gently wash the slide with distilled water from a wash bottle. Do not squirt water directly onto the smear (Figure 6.1b).

Bergey's Manual of Systematic Bacteriology, 2nd ed., 5 vols. (2000) is the standard reference for classification of prokaryotic organisms. *Bergey's Manual of Determinative Bacteriology,* 9th ed. (1994) is the standard reference for identification of culturable bacteria and archaea.

(a)   Cover the smear with carbolfuchsin for 5 minutes.

(b)   Gently wash off the carbolfuchsin with water by squirting the water so it runs through the smear.

(c)   Wash the smear with acid-alcohol for 1 minute.

(d)   Gently wash the smear with water.

(e)   Cover the smear with methylene blue for 1 minute.

(f)   Wash the smear with water.

(g)   Blot it dry.

Figure 6.1
Preparing an acid-fast stain.

4. Without drying it, wash the smear with decolorizer for 1 minute or until no more red color runs off when the slide is tipped (Figure 6.1c).

5. Wash the smear carefully with distilled water (Figure 6.1d).

6. Counterstain the smear for about 1 minute with methylene blue (Figure 6.1e).

7. Wash the smear with distilled water and blot it dry (Figure 6.1f and g).

8. Examine the acid-fast stained slide microscopically and record your observations.

9. Observe the demonstration slides.

exercise 6

Acid-Fast Staining

Name _____

Date _____

Lab Section _____

Purpose _____

_____

_____

Data

Sketch a few
bacteria (1000×).

Mycobacterium phlei

Escherichia coli

Morphology: _____   _____

Acid-fast reaction: _____   _____

Sketch a few
bacteria (1000×).

Demonstration slide 1

Demonstration slide 2

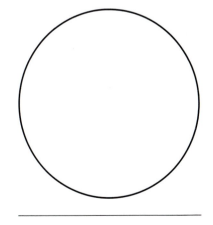

Morphology: _____   _____

Acid-fast reaction: _____   _____

Questions

1. What are the large stained areas on the sputum slide? _____

   _____

   _____

2. What is the decolorizing agent in the Gram stain? _____

   In the acid-fast stain? _____

3. What diseases are diagnosed using the acid-fast procedure? _____

   _____

   _____

4. What is phenol (carbolic acid), and what is its *usual* application? _____

   _____

   _____

Critical Thinking

1. Assuming you could stain any cell, would an acid-fast organism be gram-positive or gram-negative? Explain.

2. How do the acid-fast properties relate to the Gram stain?

3. Clinical specimens suspected of containing *Mycobacterium* are digested with sodium hydroxide (NaOH) for 30 minutes prior to staining. Why is this technique used? Why isn't this technique used for staining other bacteria?

exercise 7

Structural Stains (Endospore, Capsule, and Flagella)

Objectives

After completing this exercise, you should be able to:

1. Prepare and interpret endospore, capsule, and flagella stains.
2. Recognize the different types of flagellar arrangements.
3. Identify functions of endospores, capsules, and flagella.

Background

Structural stains can be used to identify and study the structure of bacteria. Currently, most of the fine structural details are examined using an electron microscope, but historically, staining techniques have given much insight into bacterial fine structure. We will examine a few structural stains that are still useful today. These stains are used to observe endospores, capsules, and flagella.

Endospores

Endospores are formed by several genera in the orders Bacillales, Clostridiales, and Lactobacillales.* *Bacillus* and *Clostridium* are the most familiar genera. Endospores are called "resting bodies" because they do not metabolize and are resistant to heating, various chemicals, and many harsh environmental conditions. Endospores are not for reproduction; they are formed when essential nutrients or water are not available. Once an endospore forms in a cell, the cell will disintegrate (Figure 7.1a). Endospores can remain dormant for long periods of time. However, an endospore may return to its vegetative or growing state.

Taxonomically, it is very helpful to know whether a bacterium is an endospore former and also the position of the endospores (Figure 7.1b through e). Endospores are impermeable to most stains, so heat is usually applied to drive the stain into the endospore. Once stained, the endospores do not readily decolorize. We will use the **Schaeffer–Fulton** endospore stain.

*Bergey's Manual of Systematic Bacteriology, 2nd ed. (2000).

Capsules

Many bacteria secrete chemicals that adhere to their surfaces, forming a viscous coat. This structure is called a **capsule** when it is round or oval in shape, and a **slime layer** when it is irregularly shaped and loosely bound to the bacterium. The ability to form capsules is genetically determined, but the size of the capsule is influenced by the medium on which the bacterium is growing. Most capsules are composed of polysaccharides, which are water-soluble and uncharged. Because of the capsule's nonionic nature, simple stains will not adhere to it. Most capsule-staining techniques stain the bacteria and the background, leaving the capsules unstained—essentially, a "negative" capsule stain.

Capsules have a very important role in the **virulence** (disease-causing ability) of some bacteria. For example, when bacteria such as *Streptococcus pneumoniae* have a capsule, the body's white blood cells cannot phagocytize the bacteria efficiently, and disease occurs. When *S. pneumoniae* lack a capsule, they are easily engulfed and are not virulent.

Flagella

Many bacteria are **motile,** which means they have the ability to move from one position to another in a directed manner. Most motile bacteria possess flagella, but other forms of motility occur. Myxobacteria exhibit gliding motion, and spirochetes undulate using axial filaments.

Flagella, the most common means of motility, are thin proteinaceous structures that originate in the cytoplasm and project out from the cell wall. They are very fragile and are not visible with a light microscope. They can be stained after carefully coating them using a mordant, which increases their diameter. The presence and location of flagella are helpful in the identification and classification of bacteria. Flagella are of two main types: **peritrichous** (all around the bacterium) and **polar** (at one or both ends of the cell) (Figure 7.2).

Motility may be determined by observing hanging-drop or wet mount preparations of unstained bacteria, flagella stains, or inoculation of soft (or semisolid) agar. If time does not permit doing flagella stains, observe the demonstration slides.

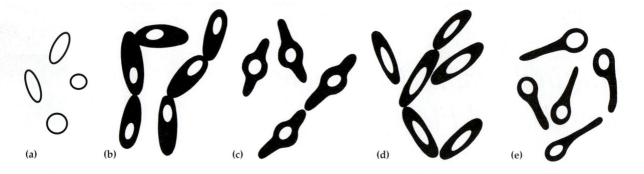

Figure 7.1

Some examples of bacterial endospores. **(a)** Free endospores, after the cell has disintegrated. **(b)** Subterminal endospores (*Bacillus macerans*). **(c)** Central, swollen endospores (*Clostridium perfringens*). **(d)** Central endospores (*Bacillus polymyxa*). **(e)** Terminal, swollen endospores (*Clostridium tetani*).

Materials

Slides

Coverslip

Paper towels

Wash bottle of distilled water

Forceps

Scalpel

Endospore stain reagents: malachite green and safranin

Capsule stain reagents: Congo red, acid-alcohol, and acid fuchsin

Flagella stain reagents: flagella mordant and carbolfuchsin

Cultures (as needed)

Endospore stain:

Bacillus megaterium (24-hour)

Bacillus subtilis (24-hour)

Bacillus subtilis (72-hour)

Capsule stain:

Streptococcus salivarius

Enterobacter aerogenes

Flagella stain:

Proteus vulgaris (18-hour)

Demonstration Slides

Endospore stain

Capsule stain

Flagella stain

Techniques Required

Compound light microscopy, Exercise 1

Smear preparation, Exercise 3

Simple staining, Exercise 3

Negative staining, Exercise 4

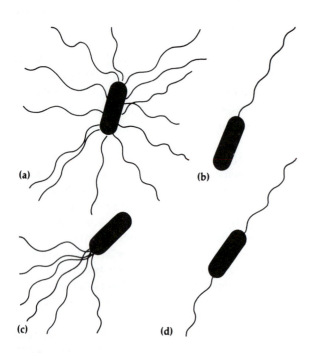

Figure 7.2

Flagellar arrangements. **(a)** Peritrichous flagella. Types of polar flagella: **(b)** monotrichous flagella, **(c)** lophotrichous flagella, and **(d)** amphitrichous flagella.

Procedure

Endospore Stain (Figure 7.3)

Be careful. The malachite green has a messy habit of ending up everywhere. But most likely you will end up with green fingers no matter how careful you are.

1. Make smears of the three *Bacillus* cultures on one or two slides, air them dry, and heat-fix them.
2. Tear out small pieces of paper towel and place them on each smear to reduce evaporation of the stain. The paper should be smaller than the slide (Figure 7.3a).
3. Flood the smears and paper with malachite green; steam the slide for 5 minutes. Add more stain as needed. *Keep it wet* (Figure 7.3b and c). What is the purpose of the paper? _____
4. Remove the towel and discard it carefully. *Do not put it in the sink.* Wash the stained smears well with distilled water (Figure 7.3d).
5. Counterstain with safranin for 30 seconds (Figure 7.3e).
6. Wash the smear with distilled water and blot it dry (Figure 7.3f).
7. Examine the slide microscopically and record your observations.
8. Observe the demonstration slides of bacterial endospores.

Capsule Stain (Figure 7.4)

1. Draw two circles on a slide. Place a loopful of Congo red in each circle (Figure 7.4a).
2. Prepare a thick smear of *S. salivarius* in the Congo red in one circle. Prepare a thick smear of *E. aerogenes* in the other circle. Let the smears air dry (Figure 7.4b and c). What color are the smears? _____
3. Fix the smear with acid-alcohol for 15 seconds (Figure 7.4d). What color is the smear? _____
4. Wash the smear with distilled water and cover it with acid fuchsin for 1 minute (Figure 7.4e and f).

5. Wash the smear with distilled water and blot it dry (Figure 7.4g).
6. Examine the slide microscopically. The bacteria will stain red, and the capsules will be colorless against a dark blue background. Record your observations.
7. Observe the demonstration slides of capsule stains.

Flagella Stain (Figure 7.5)

1. Flagella stains require special precautions to avoid damaging the flagella. Scrupulously clean slides are essential, and the culture must be handled carefully to prevent flagella from coming off the cells.
2. Without touching the bacterial culture, use a scalpel and forceps to cut out a piece of agar on which *Proteus* is growing (Figure 7.5a).

☣ Do not touch the culture with your hands.

Gently place the agar, culture-side down, on a clean glass slide (Figure 7.5b). Then carefully remove the agar with the forceps. Place the piece of agar in a Petri plate. Place the forceps and scalpel in alcohol.

3. Allow the organisms adhering to the slide to air dry. *Do not heat-fix the slide* (Figure 7.5c).
4. Cover the slide with flagella mordant and allow it to stand for 10 minutes (Figure 7.5d).
5. Gently rinse off the stain with distilled water (Figure 7.5e).
6. Cover the slide with carbolfuchsin for 5 minutes (Figure 7.5f). Rinse it gently with distilled water (Figure 7.5g). What color should the flagella be? _____ The cells? _____
7. Allow the rinsed smear to air dry (*do not blot it*), and examine it microscopically for flagella.
8. Observe the demonstration slides illustrating various flagellar arrangements. (See color plate I.3.)

(a)   Place a piece of absorbent paper over the smear.

(b)   Cover the paper with malachite green.

(c)   Steam the slide for 5 minutes.

(d)   Wash the smear with water.

(e)   Cover the smear with safranin for 30 seconds.

(f)   Wash the smear with water and blot it dry.

Figure 7.3
The endospore stain.

(a)   Place one loopful of Congo red in each circle

(b)   Prepare a thick smear of bacteria in the Congo red.

(c)   Allow the smears to air dry.

(d)   Cover each smear with acid-alcohol for 15 seconds.

(e)   Wash the smear with water.

(f)   Cover the smears with acid fuchsin for 1 minute.

(g)   Wash the smears with water and blot them dry.

Figure 7.4
The capsule stain.

(a)   Carefully cut a piece of the bacterial culture.

(b)   Gently touch the culture side of the agar piece to a clean slide.

(c)   Allow the imprint to air dry.

(d)   Cover the slide with flagella mordant for 10 minutes.

(e)   Gently rinse the slide with water.

(f)   Cover the slide with carbolfuchsin for 5 minutes.

(g)   Gently rinse the slide with water and air it dry.

Figure 7.5
The flagella stain.

exercise 7

Structural Stains (Endospore, Capsule, and Flagella)

Name _____

Date _____

Lab Section _____

Purpose _____

_____

_____

Data

Endospores

Sketch your results, and label each diagram as to color. Label the vegetative cells and endospores.

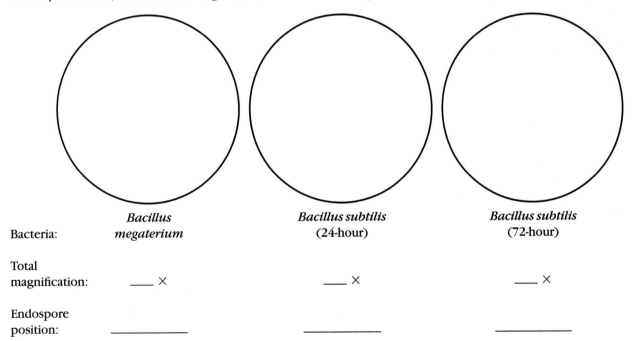

Bacteria:	*Bacillus megaterium*	*Bacillus subtilis* (24-hour)	*Bacillus subtilis* (72-hour)
Total magnification:	___ ×	___ ×	___ ×
Endospore position:	_____	_____	_____

Demonstration slides

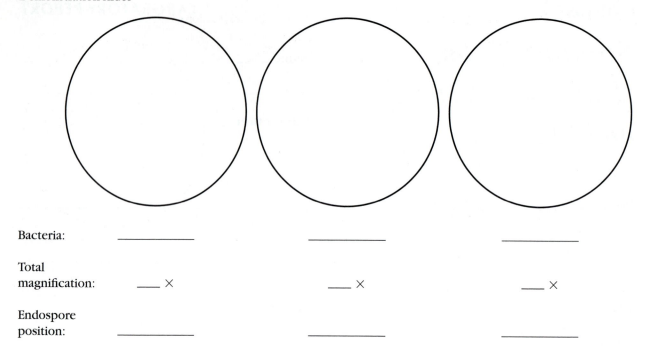

Bacteria: _____   _____   _____

Total
magnification: ___ ×   ___ ×   ___ ×

Endospore
position: _____   _____   _____

Capsules

Sketch and label the capsules and bacterial cells.   Demonstration slide

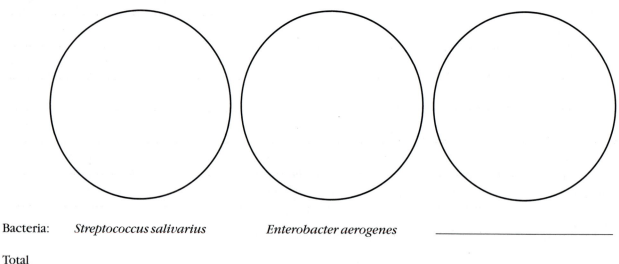

Bacteria: *Streptococcus salivarius*   *Enterobacter aerogenes*   _____

Total
magnification: ___ ×   ___ ×   ___ ×

Cell
morphology: _____   _____   _____

Flagella

Sketch and label the flagella and bacteria.   Demonstration slides

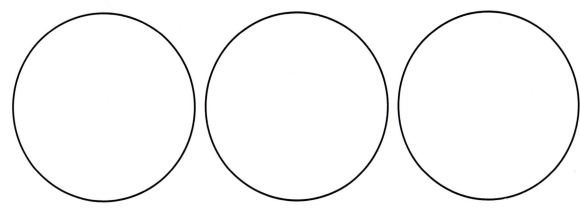

Bacteria: *Proteus vulgaris*   _____   _____

Total
magnification: ___ ×   ___ ×   ___ ×

Flagella
position: _____   _____   _____

Questions

1. What are the Gram reactions of *Clostridium* and *Bacillus*? _____

    _____

2. How might a capsule contribute to pathogenicity? _____

    How might flagella contribute to pathogenicity? _____

3. Of what advantage to *Clostridium* is an endospore? _____

    _____

    _____

4. Sketch each of the following flagellar arrangements:

    a. Monotrichous

    b. Lophotrichous

    c. Amphitrichous

    d. Peritrichous

5. How did the appearance of the 24-hour and 72-hour *Bacillus* cultures differ? How do you account for this difference? _____

_____

_____

6. Of what morphology are most bacteria possessing flagella? _____

Which morphology usually does not have flagella? _____

7. What prevents the cell from appearing green in the finished endospore stain? _____

_____

_____

Critical Thinking

1. You can see endospores by simple staining. Why not use this technique?

2. How would an endospore stain of *Mycobacterium* appear?

3. What type of culture medium would increase the size of a bacterial capsule?

4. Describe the microscopic appearance of encapsulated *Streptococcus* if stained with safranin and nigrosin.

5. In the Dorner endospore stain, a smear covered with carbolfuchsin is steamed then decolorized with acid-alcohol and counterstained with nigrosin. Describe the microscopic appearance after this procedure.

exercise 8

Morphologic Unknown

Dishonesty is knowing but ignoring the fact that the data are contradictory. Stupidity is not recognizing the contradictions.

ANONYMOUS

Objective

After completing this exercise, you should be able to identify the morphology and staining characteristics of an unknown organism.

Background

Differential staining is usually the first step in the identification of bacteria. Morphology and structural characteristics obtained from microscopic examination are also useful for identification. *Bergey's Manual** is the most widely used reference for bacterial identification. In *Bergey's Manual,* bacteria are identified by differential staining, morphology, and several other characteristics. Additional testing is needed to identify bacterial species. You will learn these techniques later in this course.

You will be given an unknown culture of bacteria. Determine its morphologic and structural characteristics. The culture contains one species (rod or coccus) and is less than 24 hours old.

Quality control (QC) is an essential component of the microbiology laboratory. The accuracy of information from stains and other tests depends on a variety of factors, including the specimen, reagents, procedures, and the person doing the test. Generally during staining, known positive and negative bacteria are stained at the same time as unknown bacteria as a check on reagents and procedures. In this exercise, you will use your results from Exercises 5 through 7 as your QC.

*Bergey's Manual of Systematic Bacteriology, 2nd ed., 5 vols. (2000) is the standard reference for classification of prokaryotic organisms. *Bergey's Manual of Determinative Bacteriology,* 9th ed. (1994) is the standard reference for identification of culturable bacteria and archaea.

Materials

Staining reagents

Culture

24-hour unknown slant culture of bacteria # _____

Techniques Required

Compound light microscopy, Exercise 1

Hanging-drop and wet mount procedures, Exercise 2

Smear preparation, Exercise 3

Simple staining, Exercise 3

Negative staining, Exercise 4

Gram staining, Exercise 5

Acid-fast staining, Exercise 6

Endospore, capsule, and flagella staining, Exercise 7

Procedure

1. Record the number of your unknown in your Laboratory Report.
2. Determine the morphology, Gram reaction, and arrangement of your unknown. Perform a Gram stain and, if needed, an endospore stain, acid-fast stain, flagella stain, wet mount technique, and capsule stain. When are the latter needed? _____

   _____
3. Tabulate your results in the Laboratory Report.

exercise 8

Morphologic Unknown

Name _____

Date _____

Lab Section _____

Results

Write *not necessary* by any category that does not apply.

Unknown # _____

Gram Stain

Controls from Exercise 5

Sketch your unknown.

Staphylococcus epidermidis	*Escherichia coli*	Unknown

Gram reaction: _____ _____ _____

Morphology:
Predominant
arrangement: _____ _____ _____

Acid-Fast Stain

Controls from Exercise 6

Sketch your unknown.

Mycobacterium phlei	*Escherichia coli*	Unknown

Acid-fast reaction: _____ _____ _____

Endospore Stain

Controls from Exercise 7

Sketch your
unknown.

24-hr *Bacillus*	72-hr *Bacillus*	Unknown

Endospores present? _____ _____ _____

Capsule Stain

Control from Exercise 7

Sketch your
unknown.

Streptococcus salivarius	Unknown

Capsules present? _____ _____

Motility

Is your unknown motile? _____

How did you determine this? _____

Questions

1. Which two stains done in this experiment are differential stains? _____

   _____

2. Assume you have Gram stained a sample of pus from a patient's urethra. You see red, nucleated cells (>10 μm) and blue rods (2.5 μm). What can you conclude? _____

   _____

   _____

Critical Thinking

1. Using *Bergey's Manual* and your textbook, place the following genera in this flowchart: *Bacillus, Corynebacterium, Escherichia, Mycobacterium, Neisseria, Sporosarcina,* and *Staphylococcus.*

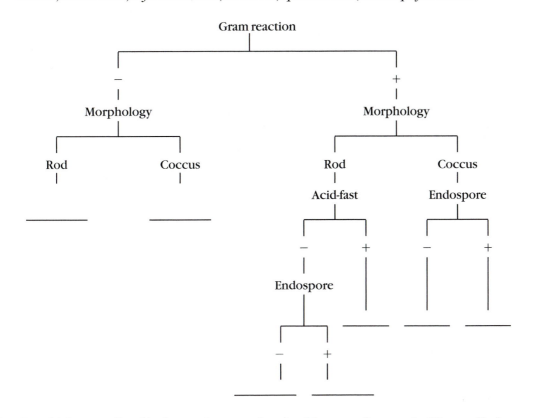

2. For which genera listed in the previous question should you test for capsules? For motility?

3. Using your textbook and this lab manual, fill in the Morphology column in the table below. Then use the information to construct a flowchart for these bacteria. Draw your flowchart on p. 66.

	Morphology	Gram Reaction	Motile	Capsule	Arrangement	Endospore
Clostridium		+	+	−	Pairs, chains	+
Enterobacter		−	+	−	Singly, pairs	−
Klebsiella		−	−	+	Singly, pairs	−
Lactobacillus		+	Rarely	−	Chains	−
Staphylococcus		+	−	−	Pairs, clusters	−
Streptococcus		+	−	Some species	Pairs, chains	−

4. Provide an example from the previous question to show that microscopic examination alone cannot be used to identify bacteria.

part three

Cultivation of Bacteria

EXERCISES

Bacteria must be grown on culture media in order to characterize and identify them. This statement from the first edition of *Bergey's Manual* summarizes the need for cultivating bacteria:

> *The earlier writers classified the bacteria solely on their morphologic characters. A more detailed classification was not possible because the biologic characters of so few of the bacteria had been determined. With the accumulation of knowledge of the biologic characters of many bacteria it was realized that it is just as incorrect to group all rods under a single genus as to group all quadruped animals under one genus.*[*]

Before attempting to culture bacteria, we must consider their nutritional requirements. Bacteria require sources of energy, carbon, nitrogen, minerals, and growth factors. Most bacteria are **chemoheterotrophs,** which require organic compounds for carbon and energy sources. Bacteria exhibit a wide range of nutritional requirements, and in many instances, enriching nutrients such as milk, serum, blood, or tomato juice must be

---

*Society of American Bacteriologists (D. H. Bergey, Chairperson). *Bergey's Manual of Determinative Bacteriology,* 1st ed. Baltimore: Williams & Wilkins, 1923, p. 1.

added for fastidious organisms that require many growth factors. **Growth factors** are organic compounds such as vitamins or amino acids that are incorporated into a cell without alteration.

The first bacterial cultures were grown in *broth* (liquid) media such as infusions and blood. Robert Koch attempted to culture bacteria on potato slices when he realized the need for *solid media*. When some bacteria wouldn't grow on a potato, he added gelatin to the liquid, but the gelatin liquefied under standard incubation conditions. Angelina Hess, a colleague of Koch, suggested the use of agar as a solidifying agent. Today, agar is the most commonly used solidifying agent in culture media.

The exercises in Part Three emphasize the use of **aseptic technique,** the prevention of unwanted microorganisms in laboratory and medical procedures.

exercise 9

Microbes in the Environment

Whatever is worth doing at all is worth doing well.

PHILIP DORMER STANHOPE

Objectives

After completing this exercise, you should be able to:

1. Describe colony morphology using accepted descriptive terms.
2. Compare bacterial growth on solid and liquid culture media.
3. Describe why agar is used in culture media.
4. Prepare nutrient broth and nutrient agar.

Background

Microbes are everywhere; they are found in the water we drink, the air we breathe, and the earth we walk on. They live in and on our bodies. Microbes occupy ecological niches on all forms of life and in most environments. In most situations, the ubiquitous microorganisms are harmless. However, in microbiology, work must be done carefully to avoid contaminating sterile media and materials with these microbes.

In this exercise, we will attempt to culture (grow) some microbes. When a medium is selected for culturing bacteria, macronutrients, an energy source, and any necessary growth factors must be provided. A medium whose exact chemical composition is known is called a **chemically defined medium** (Table 9.1).

Most chemoheterotrophic bacteria are routinely grown on **complex media**—that is, media for which the exact chemical composition varies slightly from batch to batch. Organic carbon, energy, and nitrogen sources are usually supplied by protein in the form of meat extracts and partially digested proteins called *peptones*. **Nutrient broth** is a commonly used liquid complex medium. When agar is added, it becomes a solid medium, called **nutrient agar** (Table 9.2).

Agar, an extract from marine red algae, has some unique properties that make it useful in culture media. Few microbes can degrade agar, so it remains solid during microbial growth. It liquefies at 100°C and remains in a liquid state until cooled to 40°C. Once the agar has solidified, it can be incubated at temperatures up to 100°C and remain solid.

Table 9.1
Glucose–Minimal Salts Broth

Ingredient	Amount/100 ml
Glucose	0.5 g
Sodium chloride (NaCl)	0.5 g
Ammonium dihydrogen phosphate ($NH_4H_2PO_4$)	0.1 g
Dipotassium phosphate (K_2HPO_4)	0.1 g
Magnesium sulfate ($MgSO_4$)	0.02 g
Distilled water	100 ml

Table 9.2
Nutrient Agar

Ingredient	Amount/100 ml
Peptone	0.5 g
Beef extract	0.3 g
Sodium chloride (NaCl)	0.8 g
Agar	1.5 g
Distilled water	100 ml

Media must be sterilized after preparation. The most common method of sterilizing culture media that are heat stable is **steam sterilization,** or **autoclaving,** using steam under pressure. During this process, material to be sterilized is placed in the autoclave and heated to 121°C at 15 pounds of pressure (15 psi) for 15 minutes.

Culture media can be prepared in various forms, depending on the desired use. **Petri plates** containing solid media provide a large surface area for examination of colonies. The microbes will be **inoculated,** or intentionally introduced, onto nutrient agar and into nutrient broth. The bacteria that are inoculated into culture media increase in number during an **incubation period.** After suitable incubation, liquid media become **turbid,** or cloudy, due to bacterial growth. On solid media, colonies will be visible to the naked eye. A **colony** is a population of cells that arises from a single bacterial cell. A colony may arise from a group of the same microbes attached to one another, which is therefore called a **colony-forming unit.** Although many species of bacteria give rise to colonies that appear similar, each colony that appears different is usually a different species. (See color plate VI.5.)

Materials

250 ml Erlenmeyer flask with cap or plug

100 ml graduated cylinder

Distilled water

Nutrient broth powder

Agar

Glass stirring rod

5 ml pipette

Test tubes with caps (2)

Sterile Petri dishes (4)

Balance

Weighing paper or dish

Autoclave gloves

Hot plate

Tube containing sterile cotton swabs

Tube containing sterile water

Demonstration

Use of the autoclave

Techniques Required

Pipetting, Appendix A.

Procedure

First Period

A. Preparation of culture media
    1. Prepare 100 ml of nutrient broth in a 250 ml flask. Using the graduated cylinder, add 100 ml of distilled water to the flask. Read the preparation instructions on the nutrient broth bottle. Calculate the amount of nutrient broth powder needed for 100 ml. If the amount needed for 1000 ml is _____ grams, then _____ grams are needed for 100 ml. Weigh out the required amount and add it to the flask. Stir with a glass rod until the powder is dissolved.
    2. Pipette 5 ml of the nutrient broth into each test tube and cap each tube. Label the tubes "nutrient broth." Place one in the To Be Autoclaved rack. Label the remaining tube "not sterile," and incubate it at room temperature until the next period.
    3. Add agar (1.5% w/v) to the remaining 90 ml of nutrient broth. What quantity of agar will you need to add? _____
    4. Bring the broth to a boil and continue boiling carefully until all the agar is dissolved. *Be careful:* Do not let the solution boil over. Stir often to prevent burning and boiling over.
    5. Stopper the flask, label it "nutrient agar," and place the flask and tube in the To Be Autoclaved basket.
    6. The instructor will demonstrate use of the autoclave.
    7. After autoclaving, allow the flasks and tube to cool to room temperature, or proceed to part B. What effect does the agar have on the culture medium? _____

B. Pouring plates
    Transfer the melted sterile nutrient agar flasks to a 45°C water bath. Allow the flask of nutrient agar to cool to about 45°C (warm to the touch). If the agar has solidified, it will have to be reheated to liquefy it. To what temperature will it have to be heated?

    _____

    The sterile nutrient agar must be poured into Petri plates *aseptically*—that is, without letting microbes into the nutrient medium. *Read the following procedure before beginning* so that you can work quickly and efficiently.
    1. Set four sterile, unopened Petri plates in front of you with the cover (larger half) on top. Have a lighted laboratory burner within reach on your workbench.

    > ⚠ **Keep the burner away from your hair and in the center of the bench.**

    2. Holding the flask at an angle, remove the stopper with the fourth and fifth fingers of your other hand. Heat the mouth of the flask by passing it briefly through the flame (Figure 9.1a). Why is it necessary to keep the flask at an angle through this procedure? _____
    3. Remove the cover from the first plate with the hand holding the plug. Quickly and neatly pour melted nutrient agar into the plate until

(a)   Remove the stopper, and flame the mouth of the flask.

(b)   Remove the cover from one plate, and pour nutrient agar into the plate bottom.

Figure 9.1
Petri plate pouring.

the bottom is just covered to a depth of approximately 5 mm (Figure 9.1b). Keep the flask at an angle, and replace the plate cover; move on to the next plate until all the agar is poured.

4.   When all the agar is poured, gently swirl the agar in each plate to cover any empty spaces; do not allow the agar to touch the sides or covers of the plates.

5.   To decrease condensation, leave the Petri plate covers slightly ajar for about 15 minutes until the agar solidifies.

6.   Place the empty flask in the discard area.

C.   Culturing microbes from the environment

1.   Design your own experiment. The purpose is to sample your environment and your body. Use your imagination. Here are some suggestions:

a.   You may use the lab, a washroom, or any place on campus for the environment.

b.   One nutrient agar plate might be left open to the air for 30 to 60 minutes.

c.   Inoculate a plate from an environmental surface such as the floor or workbench by

wetting a cotton swab in sterile water, swabbing the environmental surface, and then swabbing the surface of the agar. Why is the swab first moistened in sterile water? _____

_____

After inoculation, the swab should be discarded in the container of disinfectant.

2.   Inoculate two plates from the environment. Inoculate one nutrient broth tube using a swab as described in step 1c. After swabbing the agar surface, place the swab in the nutrient broth and leave it there during incubation. You may need to break off part of the wooden handle to fit the swab into the nutrient broth.

3.   The plates and tube should be incubated at the approximate temperature of the environment sampled.

4.   Inoculate two plates from your body. You could:

a.   Place a hair on the agar.

b.   Obtain an inoculum by swabbing (see step 1c) part of your body with a wet swab.

c.   Touch the plate with your fingers.

5.   Incubate bacteria from your body at or close to your body temperature. What is human body temperature? _____°C

6.   Incubate all plates inverted, so water will condense in the lid instead of on the surface of the agar. Why is condensation on the agar undesirable? _____

_____

7.   Incubate all inoculated media until the next laboratory period.

Second Period

1.   Observe and describe the resulting growth on the plates. Note each different-appearing colony, and describe the colony morphology using the characteristics given in Figure 9.2. Determine the approximate number of each type of colony. When many colonies are present, record TNTC (too numerous to count) as the number of colonies.

2.   Describe the appearance of the nutrient broth labeled "not sterile" and the broth you inoculated. Are they uniformly cloudy or turbid? Look for clumps of microbial cells, called **flocculant.** Is there a membrane, or **pellicle,** across the surface of the broth? See whether microbial cells have settled on the bottom of the tube, forming a **sediment.**

3.   Discard the plates and tubes properly.

Place contaminated plates in the biohazard bag for autoclaving.

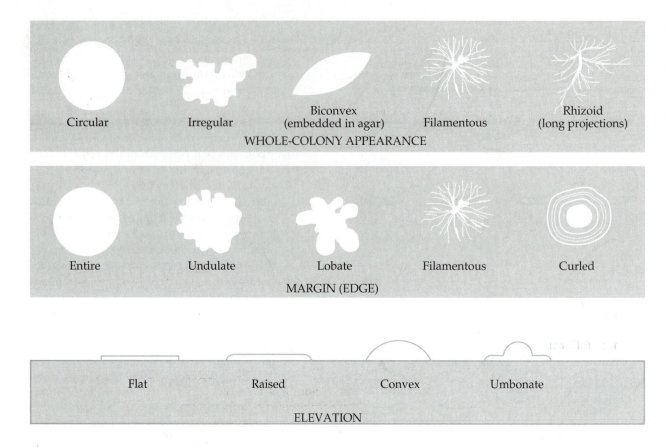

Figure 9.2
Colony descriptions.

exercise 9

Microbes in the Environment

Name _____

Date _____

Lab Section _____

Purpose _____

_____

_____

Data

Fill in the following table with descriptions of the bacterial colonies. Use a separate line for each different-appearing colony.

	Colony Description					
	Diameter	Appearance	Margin	Elevation	Color	Number of This Type
Area sampled: Sanitury Napkin Incubated at 35-37°C for 7 days		Circular	entire	flat	yellow	3
		Circular	entire	flat	white	9
Area sampled: ear piece public phone Incubated at 35-37°C for 7 days		Circular	undulate	flat/raised	yellow	6
		Irregular	undulate	flat	orange	2
	1	Irregular	undulate	flat	white cloudy	1
			filamentous			
		circular	entire	flat	white	20+
Area sampled: bathroom stall handle Incubated at 35-37°C for 7 days	2½ in	Irregular	Lobate	flat	White cloudy	2
		circular	entire	flat	peach	6

exercise 10

Transfer of Bacteria: Aseptic Technique

Study without thinking is worthless;
thinking without study is dangerous.

CONFUCIUS

Objectives

After completing this exercise, you should be able to:

1. Provide the rationale for aseptic technique.
2. Differentiate among the following: broth culture, agar slant, and agar deep.
3. Aseptically transfer bacteria from one form of culture medium to another.

Background

In the laboratory, bacteria must be cultured in order to facilitate identification and to examine their growth and metabolism. Bacteria are **inoculated,** or introduced, into various forms of culture media in order to keep them alive and to study their growth. Inoculations must be done without introducing unwanted microbes, or **contaminants,** into the media. **Aseptic technique** is used in microbiology to exclude contaminants.

All culture media are **sterilized,** or rendered free of all life, prior to use. Sterilization is usually accomplished using an autoclave. Containers of culture media, such as test tubes or Petri plates, should not be opened until you are ready to work with them, and even then, they should not be left open.

Broth cultures provide large numbers of bacteria in a small space and are easily transported. **Agar slants** are test tubes containing solid culture media that were left at an angle while the agar solidified. Agar slants, like Petri plates, provide a solid growth surface, but slants

are easier to store and transport than Petri plates. Agar is allowed to solidify in the bottom of a test tube to make an **agar deep.** Deeps are often used to grow bacteria that prefer less oxygen than is present on the surface of the medium. Semisolid agar deeps containing 0.5–0.7% agar instead of the usual 1.5% agar can be used to determine whether a bacterium is motile. Motile bacteria will move away from the point of inoculation, giving the appearance of an inverted Christmas tree.

Transfer and inoculation are usually performed with a sterile, heat-resistant, noncorroding Nichrome wire attached to an insulated handle. When the end of the wire is bent into a loop, it is called an **inoculating loop;** when straight, it is an **inoculating needle** (Figure 10.1). For special purposes, cultures may also be transferred with sterile cotton swabs, pipettes, glass rods, or syringes. These techniques will be introduced in later exercises.

Whether to use an inoculating loop or a needle depends on the form of the medium; after completing this exercise, you will be able to decide which instrument is to be used.

Materials

Tubes containing nutrient broth (3)

Tubes containing nutrient agar slants (3)

Tubes containing nutrient semisolid agar deeps (3)

Inoculating loop

Figure 10.1
(a) An inoculating loop. **(b)** A variation of the inoculating loop in which the loop is bent at a 45° angle. **(c)** An inoculating needle.

(a)   Sterilize the loop by holding the wire in the flame until it is red hot.

(b)   While holding the sterile loop and the bacterial culture, remove the cap as shown.

(c)   Briefly heat the mouth of the tube in the flame before inserting the loop for an inoculum.

(d)   Get a loopful of culture, heat the mouth of the tube, and replace the cap.

Figure 10.2
Inoculating procedures.

Inoculating needle

Test tube rack

Gram staining reagents

Cultures

Lactococcus lactis broth

Pseudomonas aeruginosa broth

Proteus vulgaris slant

Techniques Required

Compound light microscopy, Exercise 1

Smear preparation, Exercise 3

Gram staining, Exercise 5

Procedure

1. Work with only one of the bacterial cultures at a time, to prevent any mix-ups or cross-contamination.

Label one tube of each media with the name of the first culture, your name, the date, and your lab section. Inoculate each tube as described and then work with the next culture. Begin with one of the broth cultures, and gently tap the bottom of it to resuspend the sediment.

2. To inoculate nutrient broth, hold the inoculating loop in your dominant hand and one of the broth cultures of bacteria in the other hand.

    a. Sterilize the loop by holding the wire in a Bunsen burner flame (Figure 10.2a). Heat to redness. Why? _____

    b. Holding the loop like a pencil or paintbrush, curl the little finger of the same hand around the cap of the broth culture. Gently pull the cap off the tube while turning the culture tube (Figure 10.2b). If cotton stoppers are used, simply grasp the stopper with your finger. Do not set the cap down. Why not? _____

    c. Holding the tube at an angle, pass the mouth of the tube through the flame (Figure 10.2c). What is the purpose of flaming the mouth of the tube? _____
    Always hold culture tubes and uninoculated tubes at an angle to minimize the amount of dust that could fall into them. Do not tip the tube too far, or the liquid will leak out around the loose-fitting cap.

    d. Immerse the sterilized, cooled loop into the broth culture to obtain a loopful of culture (Figure 10.2d). Why must the loop be cooled first? _____
    Remove the loop, and while holding the loop, flame the mouth of the tube and recap it by turning the tube into the cap. Place the tube in your test tube rack.

    e. Remove the cap from a tube of sterile nutrient broth as previously described, and flame the mouth of the tube. Immerse the inoculating loop into the sterile broth and then withdraw it from the tube. Flame the mouth of the tube and replace the cap. Return the tube to the test tube rack.

    f. Reflame the loop until it is red and let it cool. Some prefer to hold several tubes in their hands at once (Figure 10.3). *Do not* attempt holding and transferring between multiple tubes until you have mastered aseptic transfer techniques.

3. Obtain a nutrient agar slant. Repeat steps 2a–2d, and inoculate the slant by moving the loop gently across the agar surface from the bottom of the slant to the top, being careful not to gouge the agar (Figure 10.4). Flame the mouth of the tube and replace the cap. Flame your loop and let it cool.

4. Obtain a nutrient semisolid agar deep, and using your inoculating *needle*, repeat steps 2a–2d. Inoculate the semisolid agar deep by plunging the

Figure 10.3
Experienced laboratory technicians can transfer cultures aseptically while holding multiple test tubes.

Figure 10.4
Inoculate a slant by streaking the loop back and forth across the surface of the agar.

needle straight down the middle of the deep then pulling it out through the same stab, as shown in Figure 10.5. Flame the mouth of the tube and replace the cap. Flame your needle and let it cool.

5. Using the other broth culture, label one tube of each media as described in step 1; and inoculate a broth culture, agar slant, and semisolid agar deep, as described in steps 2, 3, and 4, using your inoculating *needle*.

6. Label one tube of each media with *Proteus vulgaris* as described in step 1. To transfer *Proteus vulgaris*, flame your loop and allow it to cool. Flame the mouth of the tube, and use your inocu-

Figure 10.5
Inoculate an agar deep by stabbing into the agar with a needle.

lating loop to carefully scrape a small amount of the culture from the agar. Flame the mouth of the tube and replace the cap. Inoculate a broth and a slant as described in steps 2 and 3. Inoculate a semi-isolid agar deep with an inoculating needle as described in step 4.

7. Incubate all tubes at 35°C until the next period.
8. Record the appearance of each culture, referring to Figure 10.6.
9. Make a smear of the *Lactococcus* broth culture and the *Lactococcus* slant culture. Gram stain both smears and compare them.

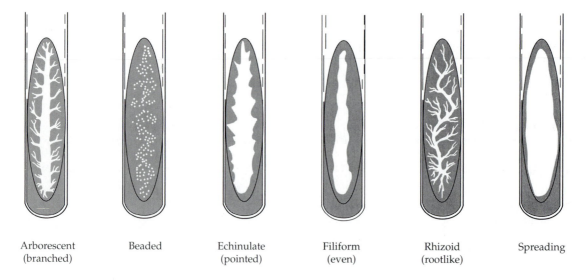

| Arborescent (branched) | Beaded | Echinulate (pointed) | Filiform (even) | Rhizoid (rootlike) | Spreading |

Figure 10.6
Patterns of growth on agar slants.

exercise 10

Transfer of Bacteria: Aseptic Technique

Name _____

Date _____

Lab Section _____

Purpose _____

_____

_____

Data

Nutrient Broth

Describe the nutrient broth cultures.

Bacterium	Is it turbid?	Is flocculant, pellicle, or sediment present?	Pigment
Lactococcus lactis			
Pseudomonas aeruginosa			
Proteus vulgaris			

Nutrient Agar Slant

Sketch the appearance of each culture. Note any pigmentation.

Lactococcus lactis

Type of growth:
_____

Pseudomonas aeruginosa

Type of growth:
_____

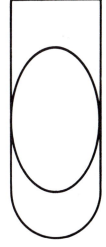
Proteus vulgaris

Type of growth:
_____

Nutrient Semisolid Agar Deep

Show the location of bacterial growth and note any pigment formation.

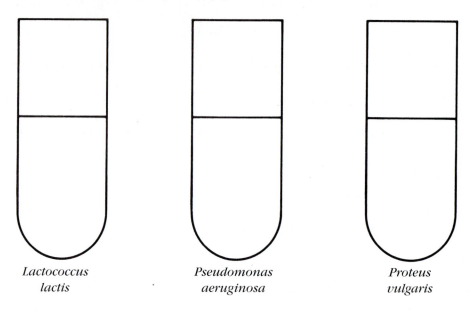

| *Lactococcus* | *Pseudomonas* | *Proteus* |
| lactis | aeruginosa | vulgaris |

Comparison of Broth and Slant Cultures

	Lactococcus lactis	
	Broth Culture	Slant Culture
Gram stain		
Morphology		
Arrangement		

Questions

1. Did growth occur at different levels in the agar deep? _____
_____
_____

2. Were any of the bacteria growing in the semisolid agar deeps motile? Explain. _____
_____
_____
_____

3. What other methods can be used to determine motility? _____
_____

4. What is the primary use of slants? Of deeps? Of broths? _____
_____
_____

5. Can you determine whether a broth culture is pure (all one species) by visually inspecting it without a microscope? _____ An agar deep culture? _____

   An agar slant culture? _____

6. When is a loop preferable for transferring bacteria? Use an illustration from your results to answer. When is a needle preferable? _____

   _____

   _____

7. What is the purpose of flaming the loop before use? After use? _____

   _____

   _____

8. Why must the loop be cool before touching it to a culture? Should you set it down to let it cool? How do you determine when it is cool? _____

   _____

   _____

9. Why is aseptic technique important? _____

   _____

   _____

Critical Thinking

1. Why was the arrangement of *Lactococcus* from the broth culture different than that from the slant culture in the second period?

2. What evolutionary advantage would there be to the formation of a pellicle in a liquid medium by a bacterium?

3. How can you tell that the media provided for this exercise were sterile?

exercise 11

Isolation of Bacteria by Dilution Techniques

Experience is the father of wisdom, and memory the mother.

THOMAS FULLER

Objectives

After completing this exercise, you should be able to:

1. Isolate bacteria by the streak plate and pour plate techniques.
2. Prepare and maintain a pure culture.

Background

In nature most microbes are found growing in environments that contain many different organisms. Unfortunately, mixed cultures are of little use in studying microorganisms because of the difficulty they present in determining which organism is responsible for any observed activity. A **pure culture,** one containing a single kind of microbe, is required in order to study concepts such as growth characteristics, pathogenicity, metabolism, and antibiotic susceptibility. Since bacteria are too small to separate directly without sophisticated micromanipulation equipment, indirect methods of separation must be used.

In the 1870s, Joseph Lister attempted to obtain pure cultures by performing serial dilutions until each of his containers theoretically contained one bacterium. However, success was very limited, and **contamination,** the presence of unwanted microorganisms, was common. In 1880 Robert Koch prepared solid media, after which microbiologists could separate bacteria by dilution and trap them on the solid media. An isolated bacterium grows into a visible colony that consists of one kind of bacterium.

Currently there are three dilution methods commonly used for the isolation of bacteria: the streak plate, the spread plate, and the pour plate. In the **streak plate technique,** a loop is used to streak the mixed sample many times over the surface of a solid culture medium in a Petri plate. Theoretically, the process of streaking the loop repeatedly over the agar surface causes the bacteria to fall off the loop one by one and ultimately to be distributed over the agar surface, where each cell develops into a colony. The streak plate is the most common isolation technique in use today.

The spread plate and pour plate are quantitative techniques that allow determination of the number of bacteria in a sample. In the **spread plate technique,** a small amount of a previously diluted specimen is spread over the surface of a solid medium using a spreading rod.

In the **pour plate technique,** a small amount of diluted sample is mixed with melted agar and poured into empty, sterile Petri dishes. After incubation, bacterial growth is visible as colonies in and on the agar of a pour plate. To determine the number of bacteria in the original sample, a plate with between 25 and 250 colonies is selected. Fewer than 25 colonies is inaccurate because a single contaminant causes at least a 4% error. A plate with greater than 250 colonies is difficult to count. The number of bacteria in the original sample is calculated using the following equation:

$$\frac{\text{Colony-forming units}}{\text{per ml}} = \frac{\text{Number of colonies}}{\text{Dilution*} \times \text{Amount plated}}$$

Materials

Petri plates containing nutrient agar (2)

Tubes containing melted nutrient agar (3)

Sterile Petri dishes (3)

250 ml beaker

Sterile 1 ml pipettes (3)

Propipette or pipette bulb

Nutrient agar slant (second period)

*In this exercise, 1 ml of sample is put into each plate. Dilution refers to the dilution of the sample (Appendix B). For example, if 37 colonies were present on the 1:8000 plate, the calculation would be as follows:

$$\text{Colony-forming units per ml} =$$

$$\frac{37}{1:8000 \times 1} = 37 \times 8000 = 296,000 = 2.96 \times 10^5$$

Cultures

Mixed broth culture of bacteria

Turbid nutrient broth from Exercise 9

Techniques Required

Compound light microscopy, Exercise 1

Aseptic technique, Exercise 10

Pipetting, Appendix A

Serial dilution techniques, Appendix B

Procedure

Streak Plate

1. Label the bottoms of two nutrient agar plates to correspond to the two broth cultures: mixed culture and turbid broth.
2. Flame the inoculating loop to redness, allow it to cool, and aseptically obtain a loopful of one broth culture.
3. The streaking procedure may be done with the Petri plate on the table (Figure 11.1a) or held in your hand (Figure 11.1b).
   a. To streak a plate (Figure 11.2), lift one edge of the Petri plate cover, and streak the first sector by making as many streaks as possible without overlapping previous streaks. Do not gouge the agar while streaking the plate. Hold the loop as you would hold a pencil or paintbrush, and gently touch the surface of the agar.
   b. Flame your loop and let it cool. Turn the plate so the next sector is on top. Streak through one area of the first sector, then streak a few times away from the first sector.
   c. Flame your loop, turn the plate again, and streak through one area of the second sector, then streak the third sector.
   d. Flame your loop, streak through one area of the third sector, then streak the remaining area of the agar surface, being careful not to make additional contact with any streaks in the previous sections. Flame your loop before setting it down. Why? _____ _____
4. Streak two plates: one of the mixed culture provided and one of the turbid broth from an environmental sample. Label each plate on the bottom with your name and lab section, the date, and the source of the inoculum.

5. Incubate the plates in an inverted position in the 35°C incubator (or at room temperature, depending on the inoculum) until discrete, isolated colonies develop (usually 24 to 48 hours). Why inverted? _____ _____
6. After incubation, record your results. Use proper terms to describe the colonies. (Refer to color plates VI.1 and VI.2.)
7. Prepare a subculture of one colony. Sterilize your needle by flaming it. Let it cool. Why use a needle instead of a loop? _____ To subculture, touch the center of a small isolated colony located on a streak line, and then aseptically streak a sterile nutrient agar slant. How can you tell whether or not you touched only one colony and whether or not you have a pure culture? _____ _____
8. Incubate the slant at 35°C until good growth is observed. Describe the growth pattern.

Pour Plate (Figure 11.3)

1. Label the bottoms of three empty, sterile Petri dishes with your name and lab section and the date. Label one plate "1:20," another, "1:400," and the third one, "1:8000." Place the labeled plates on your workbench right-side up.
2. Fill a beaker with hot (45–50°C) water (about 3–6 cm deep), and place three tubes of melted nutrient agar in the beaker. Each tube contains 19 ml of nutrient agar.
3. Select a mixed broth culture.
4. Remove a pipette, attach a bulb, and aseptically transfer 1 ml of the broth to a tube of melted agar. Mix well, as shown in Figure 11.4. Using a different pipette, transfer 1 ml from this tube to a second tube. Aseptically pour the contents of the first tube into Petri dish 1:20. Discard the pipettes in the container of disinfectant.
5. Mix the second tube. With the third pipette, aseptically transfer 1 ml to the third tube. Pour the contents of the second tube into dish 1:400. Mix the third tube and pour its contents into the remaining dish, 1:8000.
6. Discard the tubes properly. Let the agar harden in the plates, then incubate them at 35°C in an inverted position until growth is seen. *Suggestion:* When incubating multiple plates, use a rubber band to hold them together.
7. After incubation, count the number of colonies on the plates. Remember that more than 250 is too numerous to count and less than 25 is too few to count.

Figure 11.1
Inoculation of a solid medium in a Petri plate. Lift one edge of the cover while the plate **(a)** rests on the table or **(b)** is held.

Figure 11.2
Streak plate technique for pure culture isolation of bacteria. The direction of streaking is indicated by the arrows. Between each section, sterilize the loop and reinoculate with a fraction of the bacteria by going back across the previous section.

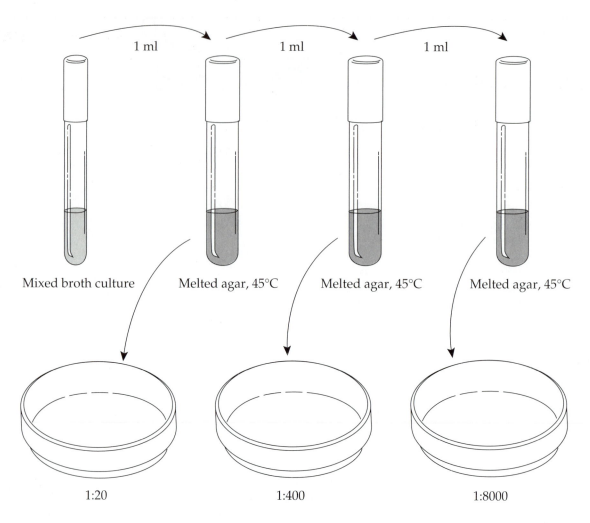

Figure 11.3
Pour plate technique. Bacteria are diluted through a series of tubes containing 19 ml of melted nutrient agar. The agar and bacteria are poured into sterile Petri dishes. The bacteria will form colonies where they are trapped in the agar.

Figure 11.4
Mix the inoculum in a tube of melted agar by rolling the tube between your hands.

exercise 11

Isolation of Bacteria by Dilution Techniques

Name _____

Date _____

Lab Section _____

Purpose _____

_____

_____

Data

Streak Plate

Sketch the appearance of the streak plates.

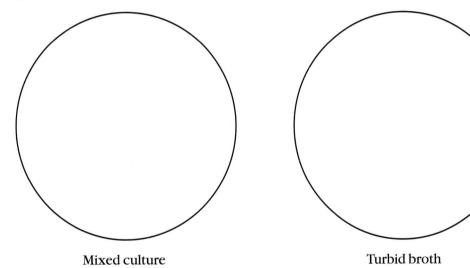

Mixed culture   Turbid broth

Fill in the following table using colonies from the most isolated streak areas.

Culture	Colony Description (Each different-appearing colony should be described.)				
	Diameter	Appearance	Margin	Elevation	Color
Mixed culture					
Turbid broth					

Pour Plate

Dilution	Number of Colonies
1:20	
1:400	
1:8000	

Calculate the number of colony-forming units per milliliter in the mixed culture. Show your calculations.

Which plate will you use for your calculations? _____

_____ Colony-forming units/ml

Subculture

Describe the growth on your slant. _____

Do you appear to have a pure culture? _____

Questions

1. How many different bacteria were in the mixed culture? _____ How many in the

   turbid broth? _____ How can you tell? _____

   _____

2. How do the colonies on the surface of the pour plate differ from those suspended in the agar? _____

   _____

   _____

3. What is a contaminant? _____

4. How would you determine whether a colony was a contaminant on a streak plate? _____

   _____

   On a pour plate? _____

5. What would happen if the plates were incubated a week longer? _____

   A month? _____

6. How could your streak plate technique be improved? _____

   _____

   _____

Critical Thinking

1. Could some bacteria grow on the streak plate and not be seen using the pour plate technique? Explain.

2. What is a disadvantage of the streak plate technique? Of the pour plate technique?

3. Will the isolated colonies always be in the fourth sector on the streak plate?

exercise 12

[handwritten: Selective media inhibit certain types of organisms]

Special Media for Isolating Bacteria

Objectives

After completing this exercise, you should be able to:

1. Differentiate between selective and differential media.
2. Provide an application for enrichment and selective media.

Background

One of the major limitations of dilution techniques used to isolate bacteria is that organisms present in limited amounts may be diluted out on plates filled with dominant bacteria. For example, if the culture to be isolated has 1 million of bacterium A and only 1 of bacterium B, bacterium B will probably be limited to the first sector in a streak plate. To help isolate organisms found in the minority, various enrichment and selective culturing methods are available that either enhance the growth of some organisms or inhibit the growth of other organisms. **Selective media** contain chemicals that prevent the growth of unwanted bacteria without inhibiting the growth of the desired organism. **Enrichment media** contain chemicals that enhance the growth of desired bacteria. Other bacteria will grow, but the growth of the desired bacteria will be increased. *[handwritten: Chocolate agar & blood agar : bacteria that need]*

Another category of media useful in identifying bacteria are **differential media.** These media contain various nutrients that allow the investigator to distinguish one bacterium from another by how they metabolize

or change the media. Differential media will be used in later exercises. *[handwritten: stool cultures]*

Because multiple methods and multiple media exist, you must be able to match the correct procedure to the desired microbe. For example, if bacterium B is salt-tolerant, salt could be added to the culture medium. Physical conditions can also be used to select for a bacterium. If bacterium B is heat-resistant, the specimen could be heated before isolation. In this exercise, we will determine one criterion used to select a culture medium. Two culture media will be compared in this exercise (Table 12.1).

Materials

Petri plate containing phenylethyl-alcohol agar

Petri plate containing tryptose agar

Gram staining reagents

Cultures

Staphylococcus epidermidis *[handwritten: Gram positive cocci cluster]*

Escherichia coli *[handwritten: gram negative-rod]*

Mixed culture of *Escherichia* and *Staphylococcus* *[handwritten: rods/coca pos & neg]*

Techniques Required

Compound light microscopy, Exercise 1

Smear preparation, Exercise 3

Table 12.1

Chemical Composition of Media Used in This Exercise

Tryptose Agar *[handwritten: Enrichment]*		Phenylethyl-Alcohol Agar *[handwritten: Selective]*	
Tryptose	1.0%	Tryptose	1.0%
Beef extract	0.3%	Beef extract	0.3%
NaCl	0.5%	Phenylethyl alcohol	0.25%
Agar	1.5%	NaCl	0.5%
		Agar	1.5%

[handwritten: one is sensitive to the alcohol]

Gram staining, Exercise 5

Aseptic technique, Exercise 10

Inoculating loop, Exercise 10

Procedure

1. Using a marker, divide each plate into three sections by labeling the bottoms. Label one section on each plate for each culture.
2. Streak each culture on the agar, as shown in Figure 12.1.
3. Incubate the plates in an inverted position at 35°C. Record the results after 24 hours, and after 48 to 72 hours, of incubation. Gram stain different-appearing colonies.

Staphylococcus epidermidis

Figure 12.1
Divide a Petri plate into three sections by drawing lines on the bottom of the plate. Inoculate each section by streaking it with an inoculating loop.

exercise 12

Special Media for Isolating Bacteria

Name _____

Date _____

Lab Section _____

Purpose _____

_____

_____

Expected Results

Before the next lab period, indicate which organism you expect to be predominant on the phenylethyl-alcohol agar, *Staphylococcus* or *Escherichia*. _____

Data

| Organism | Amount of Growth | | | | Gram Stain Results |
| | Tryptose Agar | | Phenylethyl-Alcohol Agar | | |
	24 hr	48–72 hr	24 hr	48–72 hr	
Staphylococcus epidermidis		fuzzy large colonies		sm, bubbly cloudy	pos
Escherichia coli		tiny fuzzy		sm circular bubbles	neg
mixed culture					both

Questions

1. What do your results indicate? What difference did you see between 24 and 48 to 72 hours? _____

only gram positive should grow after 24 hrs

2. Which medium is selective? Phenyl gram positive

Alcohol selective for gram positive bacteria
dissolves cell wall of gram negative

3. How did the media composition affect bacterial growth? _____

    alchol selective for

    _____

4. How did the results observed on the phenylethyl-alcohol agar plate correlate to the Gram stain? _____

    Staph grampositive grew

    _____

Critical Thinking

1. Design another medium that would select for the same group of organisms. How is your medium selective?

2. Design an enrichment medium to isolate a detergent-degrading bacterium that is found in soil.

part four

Microbial Metabolism

EXERCISES

Van Leeuwenhoek saw microorganisms in wine, and Pasteur demonstrated that the microbes were living organisms. In 1872 Pasteur wrote:

> *It is impossible that the organic matter of the newly formed ferments [microorganisms] contain a single carbon atom which has not been derived from the fermented substance.**

The chemical reactions observed by Pasteur and the chemical reactions that occur within all living organisms are referred to as **metabolism.** Metabolic processes involve **enzymes,** which are proteins that catalyze biological reactions. The majority of enzymes function inside a cell—that is, they are **endoenzymes.** Many bacteria make some enzymes, called **exoenzymes,** that are released from the cell to catalyze reactions outside of the cell (see the illustration on the next page).

Some bacteria use particular metabolic pathways in the presence of oxygen (**aerobic**) and other pathways in the absence of oxygen (**anaerobic**). Pasteur continued:

*Quoted in H. A. Lechevalier and M. Solotorovsky. *Three Centuries of Microbiology*. New York: Dover Publications, 1974, p. 26.

I have demonstrated that since this ferment [microorganism] survived in the presence of some free oxygen, it lost its fermentative abilities in proportion to the concentration of this gas.

Because many bacteria share the same colony and cell morphology, additional factors, such as metabolism, are used to characterize and classify them. On the basis of which substrates a particular bacterium uses and which metabolic products it forms, laboratory tests have been designed to determine which enzymes the bacterium has.

The first five exercises in Part Four introduce concepts in microbial metabolism and laboratory tests used to detect various metabolic activities. In Exercise 18, unknown bacteria will be identified on the basis of metabolic characteristics using commercial rapid identification methods.

Large molecules are broken down outside of a cell by exoenzymes. Smaller molecules released by this reaction are taken into the cell and further degraded by endoenzymes.

exercise 13

Carbohydrate Catabolism

The men who try to do something and fail
are infinitely better than those
who try nothing and succeed.

LLOYD JONES

simple one sugar 3-7 carbons

metabolism — break down that releases energy from decomposition of complex (carbohydrate)

Objectives

After completing this exercise, you should be able to:

1. Define the following terms: carbohydrate, catabolism, and hydrolytic enzymes.
2. Differentiate between oxidative and fermentative catabolism.
3. Perform and interpret microbial starch hydrolysis and OF tests.

Background

Chemical reactions that release energy from the decomposition of complex organic molecules are referred to as **catabolism.** Most bacteria catabolize carbohydrates for carbon and energy. **Carbohydrates** are organic molecules that contain carbon, hydrogen, and oxygen in the ratio $(CH_2O)_n$. Carbohydrates can be classified based on size: monosaccharides, oligosaccharides, and polysaccharides. **Monosaccharides** are simple sugars containing from three to seven carbon atoms. **Oligosaccharides** are composed of two to about 20 monosaccharide molecules; *disaccharides* are the most common oligosaccharides. **Polysaccharides** consist of eight or more monosaccharide molecules.

Exoenzymes are mainly **hydrolytic enzymes** that leave the cell and break down, by the addition of water, large substrates into smaller components, which can then be transported into the cell. The amylase hydrolyzes the polysaccharide starch into smaller carbohydrates. Glucose, a monosaccharide, can be released by hydrolysis (Figure 13.1). In the laboratory, the presence of an exoenzyme is determined by looking for a change in the substrate outside of a bacterial colony.

Glucose can enter a cell and be catabolized; some bacteria catabolize glucose oxidatively, producing carbon dioxide and water. **Oxidative catabolism** requires the presence of molecular oxygen (O_2). Most bacteria, however, can ferment glucose without using oxygen. **Fermentative catabolism** does not require oxygen but may occur in its presence. The metabolic

can be oxidative, fermentative or both

end-products of fermentation are small organic molecules, usually organic acids.

Whether an organism is oxidative or fermentative can be determined by using Rudolph Hugh and Einar Leifson's OF basal media with the desired carbohydrate added. **OF medium** is a nutrient semisolid agar deep containing a *high* concentration of carbohydrate and a *low* concentration of peptone. The peptone will support the growth of nonoxidative, nonfermentative bacteria. Two tubes are used: one open to the air and one sealed to keep air out. OF medium contains the indicator bromthymol blue, which turns yellow in the presence of acids, indicating catabolism of the carbohydrate. Alkaline conditions, due to the use of peptone and not the carbohydrate, are indicated by a dark blue color. If the carbohydrate is metabolized in both tubes, fermentation has occurred. Some bacteria produce gases from the fermentation of carbohydrates. An organism that can only use the carbohydrate oxidatively will produce acid in the open tube only.

Carbohydrate catabolism will be demonstrated in this exercise. These differential tests will be important in identifying bacteria in later exercises.

Figure 13.1
Starch hydrolysis. A molecule of water is used when starch is hydrolyzed.

Materials

Petri plate containing nutrient starch agar

OF-glucose deeps (2)

Mineral oil

Gram's iodine (second period)

Cultures (as assigned)

Bacillus subtilis g-positive rod

Escherichia coli g-negative rod

Pseudomonas aeruginosa g-negative rod

Alcaligenes faecalis g negative rod

Techniques Required

Inoculating loop and needle, Exercise 10

Aseptic technique, Exercise 10

Procedure

Starch Hydrolysis

1. With a marker, divide the starch agar into three sectors by labeling the bottom of the plate.
2. Streak a short single line of *Bacillus, Escherichia,* and *Pseudomonas.*
3. Incubate the plate, inverted, at 35°C for 24 hours. After growth occurs, the plate may be refrigerated at 5°C until the next lab period.
4. Record any bacterial growth, then flood the plate with Gram's iodine (Figure 13.2). Areas of starch hydrolysis will appear clear, while unchanged starch will stain dark blue. (See color plate V.5.) Record your results.

Figure 13.2
Starch hydrolysis test. After incubation, add iodine to the plate to detect the presence of starch.

OF-Glucose

1. Using an inoculating needle, inoculate two tubes of OF-glucose media with the assigned bacterial culture (*Escherichia, Pseudomonas,* or *Alcaligenes*).
2. Pour about 5 mm of mineral oil over the medium in one of the tubes. Replace the cap.
3. Incubate both tubes at 35°C until the next lab period.
4. Compare the inoculated tubes and an uninoculated OF-glucose tube. Record the following: the presence of growth, whether glucose was used, and the type of metabolism. Motility can also be ascertained from the OF tubes. How? _____

_____

5. Observe and record the results from the microorganisms you did not culture. (See color plate III.1.)

exercise 13

Carbohydrate Catabolism

Name _____

Date _____

Lab Section _____

Purpose _____

_____

_____

Data

Record your results in the following tables.

Starch Hydrolysis

Organism	Growth	Color of Medium Around Colonies After Addition of Iodine	Starch Hydrolysis: (+) = yes (−) = no
Bacillus subtilis			
Pseudomonas aeruginosa			
Escherichia coli			

OF-Glucose

Color of uninoculated medium: _____

Organism	Growth		Color		Fermenter (F), Oxidizer (O), Neither (−)	Motile
	Open Tube	Plugged Tube	Open Tube	Plugged Tube		
Pseudomonas aeruginosa						
Alcaligenes faecalis		No growth		dark green		
Escherichia coli						

Questions

1. Which organism(s) gave a positive test for starch hydrolysis? _____

    _____

    How can you tell? _____

2. What would be found in the clear area that would not be found in the blue area of a starch agar plate after the

    addition of iodine? _____

    _____

3. How can you tell amylase is an exoenzyme and not an endoenzyme? _____

    _____

4. How can you tell from OF-glucose medium whether an organism oxidizes glucose? _____

    _____

    Ferments glucose? _____

    Doesn't use glucose? _____

5. If an organism grows in the OF-glucose medium that is exposed to air, is the organism oxidizing or fermenting

    glucose? Explain. _____

    _____

6. Why is it important to first determine if growth occurred in a differential medium, such as starch agar, before

    examining the plate for starch hydrolysis? _____

    _____

    _____

Critical Thinking

1. Some microbes metabolized the peptone in the OF-glucose medium, and the indicator turned blue, indicating
    alkaline products. What alkaline product was produced?

2. How can organisms that don't use starch grow on a starch agar plate?

3. If iodine were not available, how would you determine if starch hydrolysis had occurred?

4. Locate the H$^+$ and OH$^-$ ions from the water molecule that was split (hydrolyzed) in this reaction, showing esculin hydrolysis.

Esculin $+ H_2O \longrightarrow$ Glucose $+$ Esculetin

exercise 14

Fermentation of Carbohydrates

EXERCISE

14

Objectives

After completing this exercise, you should be able to:

1. Define fermentation.
2. Perform and interpret carbohydrate fermentation tests.
3. Perform and interpret the MR and V–P tests.

Background

Once a bacterium has been determined to be fermentative by the OF test, further tests can determine which carbohydrates, in addition to glucose, are fermented; in some instances, the end-products can also be determined. Many carbohydrates—including monosaccharides such as glucose, disaccharides like sucrose, and polysaccharides such as cellulose—can be fermented. Many bacteria produce organic acids (for example, lactic acid) and hydrogen and carbon dioxide gases from carbohydrate fermentation. A **fermentation tube** is used to detect acid and gas production from carbohydrates. The fermentation medium contains peptone, an acid–base indicator (phenol red), an inverted tube to trap gas, and 0.5–1.0% of the desired carbohydrate. In Figure 14.1, phenol red indicator is red (neutral) in an uninoculated fermentation tube; fermentation that results in acid production will turn the indicator yellow (pH of 6.8 or below). When gas is produced during fermentation, some will be trapped in the inverted tube. Fermentation occurs with or without oxygen present; however, during prolonged incubation periods (greater than 24 hours), many bacteria will begin growing oxidatively on the peptone after exhausting the carbohydrate supplied, causing neutralization of the indicator and turning it red.

Fermentation processes can produce a variety of end-products, depending on the substrate, the incubation, and the organism. In some instances, large amounts of acid may be produced, while in others, a majority of neutral products may result (Figure 14.2a). The **MRVP test** is used to distinguish between organisms that produce large amounts of acid from glucose and those that produce the neutral product *acetoin*. MRVP medium is a glucose-supplemented nutrient broth used for the **methyl red (MR) test** and the **Voges–Proskauer (V–P) test.** If an organism produces a large amount of organic acid from glucose, the medium will remain red when methyl red is added, indicating that the pH is below 4.4. If neutral products

Figure 14.1

Carbohydrate fermentation tube. **(a)** The phenol red indicator is red in a neutral or alkaline solution. **(b)** Phenol red turns yellow in the presence of acids. **(c)** Gases are trapped in the inverted tube while the indicator shows the production of acid.

are produced, methyl red will turn yellow, indicating a pH above 6.0. The production of acetoin is detected by the addition of potassium hydroxide and α-naphthol. If acetoin is present, the upper part of the medium will turn red; a negative V–P test will turn the medium light brown. The chemical process is shown in Figure 14.2b. The production of acetoin is dependent on the length of incubation, as shown in Figure 14.3.

Materials

Glucose fermentation tube

Lactose fermentation tube

Sucrose fermentation tube

MRVP broths (4)

Second period:

MRVP broth

Parafilm squares

Empty test tube

(a)

Glucose — Embden–Meyerhof pathway (glycolysis) → Pyruvic acid

Butylene glycol pathway → Lactic acid

Acetoin (acetylmethylcarbinol) + 2 CO_2

(b)

Acetoin + α-Naphthol (catalyst) + KOH + O_2 → Diacetyl + Guanidine group (in peptones under alkaline conditions) — Condensation → Pinkish-red product

Figure 14.2

MRVP test. **(a)** Organic acids, such as lactic acid, or neutral products, such as acetoin, may result from fermentation. **(b)** Potassium hydroxide (KOH) and α-naphthol are used to detect acetoin.

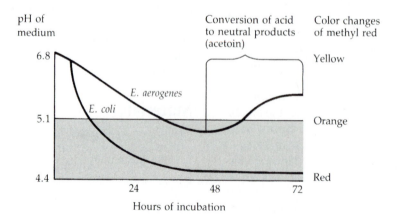

Figure 14.3

The production of acetoin is dependent on incubation time and pH.

Methyl red

V–P reagent I, α-naphthol solution

V–P reagent II, potassium hydroxide (40%)

Cultures (as assigned)

Escherichia coli

Enterobacter aerogenes

Alcaligenes faecalis

Proteus vulgaris

Techniques Required

Inoculating loop, Exercise 10

Aseptic technique, Exercise 10

OF test, Exercise 13

Procedure

Fermentation Tubes

1. Use a loop to inoculate the fermentation tubes with the assigned bacterial culture.
2. Incubate the tubes at 35°C. Examine them at 24 and 48 hours for growth, acid, and gas production. Compare them to an uninoculated fermentation tube. Why is it important to record the presence of

growth? _____

3. Record your results with this culture, as well as your results for the other species tested. (See color plate III.2.)

MRVP Tests

1. Using a loop, inoculate two MRVP tubes with *Escherichia* and two with *Enterobacter*.
2. Incubate the tubes at 35°C for 48 hours or longer. Why is time of incubation important? _____

_____

3. To one tube of each set (*Escherichia* and *Enterobacter*), add 5 drops of methyl red. Record the resulting color. Red is a positive methyl red test.
4. To the other set of two tubes, add 0.6 ml (12 drops) of V–P reagent I and 0.2 ml (2 or 3 drops) of V–P reagent II.
5. Cover each tube with a Parafilm square, and shake the tubes.
6. Leave the caps off to expose the media to oxygen in order to oxidize the acetoin (Figure 14.2b). Allow the tubes to stand for 15 to 30 minutes. A positive V–P test will develop a pinkish-red color.
7. Controls. Pour half of the contents of an uninoculated MRVP broth into an empty test tube. Perform the MR test (step 3) on one tube. Perform the V–P test (steps 4–6) on the other tube.
8. Record your results. (See color plates III.8 and III.9.)

exercise 14

LABORATORY REPORT

Fermentation of Carbohydrates

Name _____

Date _____

Lab Section _____

Purpose _____

_____

_____

Data

Fermentation Tubes

Color of uninoculated medium: _____

If yellow then acid was produced (handwritten)

Organism		Growth	Color	Acid	Gas	Growth	Color	Acid	Gas	Growth	Color	Acid	Gas
			Glucose				Lactose				Sucrose		
Escherichia coli	24 hr (7 days)	yes chunky	yellow	yes	yes	cloudy	opay	no	yes	yes white chunky	pinkish	no	yes
	48 hr												
Enterobacter aerogenes	24 hr		some acid +					yes				some acid +-	
	48 hr												
Alcaligenes faecalis	24 hr												
	48 hr												
Proteus vulgaris	24 hr												
	48 hr												

(handwritten annotations in left margin: *gram rod*, *gram rod*, *gram rod*, *gram rod*)

MRVP Tests

handwritten note at top: 5 drops methal red (red is positive) 12 drops VP I   2-3 drops VP II

Organism	Growth	MR		V-P	
		Color	+ or −	Color	+ or −
Escherichia coli	*cloudy yellowish*	*reddish pinkish*	*positive*	*yellow orange*	*negative*
Enterobacter aerogenes					
Controls	—		—		—

Questions

1.   Why are fermentation tubes evaluated at 24 and 48 hours? _____

_____

What would happen if an organism used up all the carbohydrate in a fermentation tube? _____

_____ What would the

organism use for energy? _____ What color would the indicator be then? _____

2.   If an organism oxidatively metabolizes glucose, what result will occur in the fermentation tubes? _____

_____

_____

3.   Were these media differential or selective? _____

Critical Thinking

1.   Could an organism be both MR and V–P positive? Explain.

2.   Could an organism be a fermenter and also be both MR and V–P negative? Explain.

3. How could you determine whether a bacterium fermented the following carbohydrates: mannitol, sorbitol, adonitol, or arabinose?

4. If a bacterium cannot ferment glucose, why not test its ability to ferment other carbohydrates?

cover each tube w/ a square & shake tubes
wait 15-20 min
pinC positive

exercise 15

Protein Catabolism, Part 1

Objectives

After completing this exercise, you should be able to:

1. Determine a bacterium's ability to hydrolyze gelatin.
2. Test for the presence of urease.
3. Perform and interpret a litmus milk test.

Background

Proteins are organic molecules that contain carbon, hydrogen, oxygen, and nitrogen; some proteins also contain sulfur. The subunits that make up a protein are called **amino acids** (Figure 15.1). Amino acids bond together by **peptide bonds** (Figure 15.2), forming a small chain (a **peptide**) or a larger molecule (a **polypeptide**).

Bacteria can hydrolyze the peptides or polypeptides to release amino acids (Figure 15.2). They use the amino acids as carbon and energy sources when carbohydrates are not available. However, amino acids are primarily used in anabolic reactions.

Large protein molecules, such as gelatin, are hydrolyzed by exoenzymes, and the smaller products of hydrolysis are transported into the cell. Hydrolysis of gelatin can be demonstrated by growing bacteria in nutrient gelatin. Nutrient gelatin dissolves in warm water (50°C), solidifies (**gels**) when cooled below 25°C, and liquefies (**sols**) when heated to about 25°C. When an exoenzyme hydrolyzes gelatin, it liquefies and does not solidify even when cooled below 20°C.

Some bacteria can hydrolyze the protein in milk called **casein.** Casein hydrolysis can be detected in litmus milk. Litmus milk consists of skim milk and the indicator litmus. The medium is opaque, due to casein in colloidal suspension, and the litmus is blue. After **peptonization** (hydrolysis of the milk proteins), the medium becomes clear as a result of hydrolysis of casein to soluble amino acids and peptide fragments. **Catabolism of amino acids** will result in an alkaline (purple) reaction. Litmus milk is also used to detect **lactose fermentation,** since litmus turns pink in the presence of acid. Excessive amounts of acid will cause **coagulation** (curd formation) of the milk. In addition, some bacteria can **reduce litmus,** causing the litmus indicator to turn white in the bottom of the tube.

Urea is a waste product of protein digestion in most vertebrates and is excreted in the urine. Presence of the enzyme **urease,** which liberates ammonia from urea (Figure 15.3), is a useful diagnostic test for identifying bacteria. **Urea agar** contains peptone, glucose, urea,

Figure 15.1

General structural formula for an amino acid. The letter R stands for any of a number of groups of atoms. Different amino acids have different R groups.

Figure 15.2

Hydrolysis of peptides. A molecule of water is used when the peptide is hydrolyzed.

Peptide bond

Aspartame (Nutrasweet®), a dipeptide

Methylated phenylalanine

Aspartic acid

$$H_2N$$
$$\underset{H_2N}{\overset{H_2N}{>}}C=O + H_2O \xrightarrow{\text{Urease}} 2\,NH_3 + CO_2$$

H₂N — Urea   Water   Ammonia   Carbon dioxide

Figure 15.3
Urea hydrolysis.

and phenol red. The pH of the prepared medium is 6.8 (phenol red turns yellow). During incubation, bacteria possessing urease will produce ammonia, which raises the pH of the medium, turning the indicator fuchsia (hot pink) at pH 8.4.

We will investigate bacterial action on nutrient gelatin, litmus milk, and urea agar in this exercise, and amino acid metabolism in Protein Catabolism, Part 2.

Materials

Tubes containing nutrient gelatin (2)

Tubes containing litmus milk (2)

Tubes containing urea agar (2)

Cultures

gram rod

Pseudomonas aeruginosa

gram rod

Proteus vulgaris ~ very motile

Techniques Required

Inoculating loop, Exercise 10

Aseptic technique, Exercise 10

Procedure

1. Label one tube of each medium *"Pseudomonas"* and the other tube *"Proteus."*

2. Gelatin hydrolysis. Examine the nutrient gelatin: Is it solid or liquid? _____
    What is the temperature of the laboratory? _____
    If the gelatin is solid, what would you need to do to liquefy it? _____
    To resolidify it? _____
    a. Inoculate one tube with *Pseudomonas* and one with *Proteus*, using your inoculating needle.
    b. Incubate the tubes at room temperature, and record your observations at 2 to 4 days and again at 4 to 7 days. Do not agitate the tube when the gelatin is liquid. Why? _____
    c. If the gelatin has liquefied, place the tube in a beaker of crushed ice for a few minutes. Is the gelatin still liquefied? _____
    Record your results. Indicate liquefaction or hydrolysis by (+). (See color plate III.6.)

3. Litmus milk. Describe the appearance of litmus milk. _____
    _____
    a. Inoculate one tube with *Pseudomonas* and one with *Proteus*.
    b. Incubate the litmus milk at 35°C for 24 to 48 hours and record the results. Litmus is pink under acidic conditions, purple under alkaline conditions, and white when reduced. (See color plate III.3.) Compare your results with an uninoculated litmus milk tube.

4. Urease test.
    a. Inoculate one urea agar slant with *Pseudomonas* and one with *Proteus*.
    b. Incubate the tubes for 24 to 48 hours at 35°C. (See color plate III.13.) Record your results: (+) for the presence of urease (red) and (−) for no urease. What color is phenol red at pH 6.8 or below? _____
    At pH 8.4 or above? _____

exercise 15

Protein Catabolism, Part 1

Name _____

Date _____

Lab Section _____

Purpose _____

_____

_____

Data

Fill in this table and the one on the next page.

Controls	Appearance of Uninoculated Tube
Pseudomonas Gelatin *Proteus*	liquidfyed slightly cloudy has little tiny particles skin like texture little particle floating
Pseudomonas Litmus milk *Proteus*	very thick settle out at the bottom a little purple clear fluid layer very white thick cloudy chunky seperated clear fluid layer
Pseudo Urea agar *Proteus*	yellowish pink cloudy fuschia

Diagram the appearance of the gelatin.

Pseudomonas aeruginosa

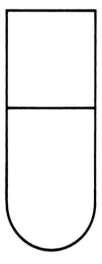

Proteus vulgaris

Test	Results	
	Pseudomonas aeruginosa	*Proteus vulgaris*
Gelatin hydrolysis at ___7___ days Growth?	Cloudy	Skin like texture
Hydrolysis?	yes/after ice. still hydrolysis	yes hydrolysis
Gelatin hydrolysis at _____ days Growth?		
Hydrolysis?		
Litmus milk Peptonization		
Acid		
Alkaline		
Coagulation	yes very thick	yes very thick
Reduction		
No change		
Urea agar Growth?	yes white	yes white
Hydrolysis?	not positive still yellowish	positive ammonia one turns indicator fuschia

Questions

1. Nutrient gelatin can be incubated at 35°C. What would have to be done to determine hydrolysis after incubation

    at 35°C? _____

    _____

    _____

2.   What is a disadvantage of litmus milk medium in diagnostic tests? _____

_____

_____

3.   What happens to milk when the suspended (colloidal) proteins are hydrolyzed? _____

_____

_____

4.   What is the source of urea in the body? _____

_____

_____

Critical Thinking

1.   Why is agar used as a solidifying agent instead of gelatin?

2.   What would you find in the liquid of hydrolyzed gelatin?

3.   When changing a baby's wet diapers, one smells ammonia. Why?

Can the organism produce iron

exercise 16

Protein Catabolism, Part 2

Objectives

After completing this exercise, you should be able to:

1. Define the following terms: deamination and decarboxylation.
2. Explain the derivation of H_2S in decomposition.
3. Perform and interpret an indole test.
4. Interpret results from MIO medium.

Background

Once amino acids are taken into a bacterial cell, various metabolic processes can occur. Before an amino acid can be used as a carbon and energy source, the amino group must be removed. The removal of an amino group is called **deamination.** The amino group is converted to ammonia, which can be excreted from the cell.

Deamination results in the formation of an organic acid. Deamination of the amino acid phenylalanine can be detected by forming a colored ferric ion complex with the resulting acid (Figure 16.1). Deamination can also be ascertained by testing for the presence of ammonia using *Nessler's reagent,* which turns deep yellow in the presence of ammonia.

Various amino acids may be decarboxylated. **Decarboxylation** is the removal of carbon dioxide from an amino acid. The presence of a specific decarboxylase enzyme results in the breakdown of the amino acid with the formation of the corresponding amine, liberation of carbon dioxide, and a shift in pH to alkaline. Media for decarboxylase reactions consist of glucose, nutrient broth, a pH indicator, and the desired amino acid. In Figure 16.2, bromcresol purple is used as a pH indicator, and a positive decarboxylase test yielding

L-Phenylalanine

Phenylalanine deaminase

Phenylpyruvic acid + Ammonia

Phenylpyruvic acid + Ferric ion (Fe^{3+}) $\longrightarrow$ Green complex

Figure 16.1
Phenylpyruvic acid resulting from the deamination of phenylalanine is detected by the addition of ferric ion (Fe^{3+}).

motility
use needle

Ornithine decarboxylase

Ornithine
amino acid
Bromcresol purple
(yellow)

Putrescine

Bromcresol purple
(lavender-purple)

Figure 16.2
Decarboxylation of the amino acid ornithine causes a change in the bromcresol purple indicator.

$$HSCH_2CH(NH_2)COOH \xrightarrow{\text{Cysteine desulfhydrase}} H_2S + NH_3 + CH_3COCOOH$$

Cysteine   Pyruvic acid

$$H_2S \quad + \quad FeSO_4 \longrightarrow FeS \quad + \quad H_2SO_4$$

Hydrogen sulfide   Ferrous sulfate   Ferrous sulfide   Sulfuric acid
(black precipitate)

Figure 16.3

If iron was produced

The release of H_2S from the amino acid cysteine is detected by the formation of ferrous sulfide.

Tryptophan   Indole   Pyruvic acid

Indole   Kovacs reagent (*next week*)
(*p*-Dimethylamino-
benzaldehyde)

Rosindole dye
(bright red)

acid concentration

Figure 16.4

Many bacteria produce indole from the amino acid tryptophan.

excess amines is indicated by purple. (Bromcresol purple is yellow in acidic conditions.) The names given to some of the amines, such as putrescine, indicate how foul-smelling they are. Cadaverine was the name given to the foul-smelling diamine derived from decarboxylation of lysine in decomposing bodies on a battlefield.

Some bacteria liberate **hydrogen sulfide (H_2S)** from the sulfur-containing amino acids: cystine, cysteine, and methionine. H_2S can also be produced from the reduction of inorganic compounds, such as thiosulfate ($S_2O_3^{2-}$). H_2S is commonly called rotten egg gas because of the copious amounts liberated when eggs decompose. To detect H_2S production, a heavy-metal salt containing ferrous ion (Fe^{2+}) is added to a nutrient

culture medium. When H_2S is produced, the sulfide (S^{2-}) reacts with the metal salt to produce a visible black precipitate. The production of hydrogen sulfide from cysteine is shown in Figure 16.3.

The ability of some bacteria to convert the amino acid tryptophan to indole or a blue compound called indigo is a useful diagnostic tool (Figure 16.4). The **indole test** is performed by inoculating a bacterium into tryptone broth and detecting indole by the addition of dimethylaminobenzaldehyde (**Kovacs reagent**). In this exercise, a differential screening medium, motility indole ornithine (MIO) agar, will be used. **MIO** is a single culture medium in which motility, indole production, and ornithine decarboxylase activity can be determined.

Materials

Phenylalanine slants (2)

Peptone iron deeps (2)

MIO deeps (2)

Second period:

Ferric chloride reagent

Kovacs reagent

Cultures (as needed)

Escherichia coli   ~~Pepton Iron~~ *Phenylalanine*

Pseudomonas aeruginosa

Proteus vulgaris   *Phenylalanine*
Enterobacter aerogenes   *Peptone Iron*

Techniques Required

Inoculating loop, Exercise 10

Aseptic technique, Exercise 10

Procedure

Phenylalanine Deamination

1. Streak one phenylalanine slant heavily with *Proteus vulgaris* and the other with one of the remaining bacterial cultures.
2. Incubate the tubes for 1 to 2 days at 35°C. Observe for the presence of growth.

3. Add 4 or 5 drops of ferric chloride reagent to the top of the slant, allowing the reagent to run through the growth on the slant. A positive test gives a dark green color. (See color plate III.15.)

Hydrogen Sulfide Production

1. Stab one peptone iron deep with *Escherichia coli* and the other with *Proteus vulgaris*.
2. Incubate the tubes at 35°C for up to 7 days. Observe initially at 24 or 48 hours.
3. Observe for the presence of growth. Blackening in the butt of the tube indicates a positive test. (See color plate III.4.)

MIO

1. Stab one MIO deep with *Enterobacter aerogenes* and the other with *Proteus vulgaris*.
2. Incubate the tubes for 24 hours at 35°C.
3. Compare the inoculated tube with an uninoculated tube. Observe for the presence of growth. Motility is demonstrated by growth diffusing out from the stab inoculation line or by clouding of the medium.
4. The ornithine decarboxylation reaction is indicated by a purple color; a negative is yellow. Why?

_____

_____

5. Add 4 to 5 drops of Kovacs reagent. Mix the tube gently. A cherry red color indicates a positive test. (See color plate III.14.)

Results

Record your results for each test in the Laboratory Report.

Break down of protein

exercise 16

Protein Catabolism, Part 2

Name _____

Date _____

Lab Section _____

Purpose _____

_____

_____

Data

Fill in the following table.

Organism	Test						MIO					
	Phenylalanine deaminase			Hydrogen sulfide			Motility	Indole			Ornithine decarboxylase	
	Original color: _____			Original color: _____			Original color: _____					
	Growth	Color	Reaction (+ or −)	Growth	Color	Reaction (+ or −)	Growth	Motility (+ or −)	Color	Reaction (+ or −)	Color	Reaction (+ or −)
Escherichia coli							Not tested	Not tested			Not tested	
Pseudomonas aeruginosa				Not tested			Not tested	Not tested			Not tested	
Proteus vulgaris												
Enterobacter aerogenes				Not tested								

Questions

1. When spoilage of canned foods occurs, what causes the blackening of the cans? _____

_____

_____

2. Show how lysine could be decarboxylated to give the end-products indicated.

$$H_2N-\overset{\overset{\displaystyle H}{|}}{\underset{\underset{\displaystyle H}{|}}{C}}-\overset{\overset{\displaystyle H}{|}}{\underset{\underset{\displaystyle H}{|}}{C}}-\overset{\overset{\displaystyle H}{|}}{\underset{\underset{\displaystyle H}{|}}{C}}-\overset{\overset{\displaystyle H}{|}}{\underset{\underset{\displaystyle H}{|}}{C}}-\overset{\overset{\displaystyle H}{|}}{\underset{\underset{\displaystyle NH_2}{|}}{C}}-COOH \rightarrow$$

Lysine

$+$

Cadaverine

_____

3. Show how alanine could be deaminated to give the end-products indicated.

$$H_3N-\overset{\overset{\displaystyle CH_3}{|}}{\underset{\underset{\displaystyle H}{|}}{C}}-COOH \longrightarrow$$

Alanine

Pyruvic acid

$+$

_____

Of what value is deamination to a microbe? _____

_____

Critical Thinking

1. Why look for black precipitate (FeS) in the butt instead of on the surface of an H_2S test?

2. Decarboxylation of an amino acid results in the evolution of carbon dioxide. Would a gas trap, such as that used in a fermentation test, be an accurate measure of decarboxylation?

3. In blue-diaper syndrome, a baby's urine turns blue in its diapers following the administration of oral tryptophan. The baby's serum and urine levels of tryptophan are very low. What is causing the blue pigment to appear?

exercise 17

Respiration

Objectives

After completing this exercise, you should be able to:

1. Define reduction.
2. Compare and contrast the following terms: aerobic respiration, anaerobic respiration, and fermentation.
3. Perform nitrate reduction, catalase, and oxidase tests.

Background

Molecules that combine with electrons liberated during metabolic processes are called **electron acceptors.** Electron acceptors become **reduced** when they gain electrons. Electrons are formed from the ionization of a hydrogen atom, as shown here:

$$H \longrightarrow H^+ \quad + \quad e^-$$

Hydrogen atom   Hydrogen ion   Electron

When an electron acceptor picks up an electron, it becomes negatively charged and combines with the positively charged hydrogen ion. **Reduction,** then, is a gain of electrons or hydrogen atoms.

Organic molecules act as electron acceptors in fermentative metabolism. Inorganic molecules serve as electron acceptors in oxidative metabolism or **respiration.** Molecular oxygen (O_2) is the final electron acceptor in **aerobic respiration.**

In aerobic bacteria, cytochromes carry electrons to O_2. Four general classes of bacterial cytochromes have been identified, and the **oxidase test** is used to determine the presence of one of these, cytochrome *c*. The oxidase test is useful in identifying bacteria.

During aerobic respiration, hydrogen atoms may combine with oxygen, forming hydrogen peroxide (H_2O_2), which is lethal to the cell. Most aerobic organisms produce the enzyme **catalase,** which breaks down hydrogen peroxide to water and oxygen, as shown here:

$$2\,H_2O_2 \xrightarrow{\text{Catalase}} 2\,H_2O \;+\; O_2\uparrow$$

Hydrogen peroxide   Water   Oxygen

In the process of **anaerobic respiration,** inorganic compounds other than O_2 act as final electron acceptors. (A few bacteria use organic electron acceptors in anaerobic respiration.) During anaerobic respiration, some bacteria reduce nitrates to nitrites; others further reduce nitrites to nitrous oxide or nitrogen gas.

Nitrate broth (nutrient broth plus 0.1% potassium nitrate) is used to determine a bacterium's ability to reduce nitrates. Nitrites are detected by the addition of dimethyl-α-naphthylamine and sulfanilic acid to nitrate broth. A red color, indicating that nitrite is present, is a positive test for nitrate reduction. A negative test (no nitrites) is further checked for the presence of nitrate in the broth by the addition of zinc. If nitrates are present, reduction has not taken place. Zinc will reduce nitrate to nitrite, and a red color will appear. If neither nitrate nor nitrite is present, the nitrogen has been reduced to nitrous oxide (N_2O) or nitrogen gas (N_2).

The reduction of some inorganic compounds in respiration is shown in Figure 17.1. In this exercise, we will examine the oxidase test, the catalase test, and the reduction of nitrate.

Materials

Tubes containing nitrate broth (3)

Petri plates containing trypticase soy agar (2)

Second period:

Tube containing nitrate broth

Nitrate reagent A (dimethyl-α-naphthylamine)

Nitrate reagent B (sulfanilic acid)

Hydrogen peroxide, 3%

Sterile toothpick

Oxidase reagent or oxidase disk

Zinc dust

Cultures

Bacillus subtilis

Escherichia coli

Pseudomonas aeruginosa

Lactococcus lactis

Techniques Required

Inoculating loop, Exercise 10

Aseptic technique, Exercise 10

Aerobic respiration:

$$\tfrac{1}{2} O_2 + 2\,H^+ + 2\,e^- \longrightarrow H_2O$$

Oxygen   Water

Anaerobic respiration:

$$SO_4^{2-} + 10\,H^+ + 10\,e^- \longrightarrow H_2S + 4\,H_2O$$

Sulfate ion   Hydrogen
                                       sulfide

$$CO_3^{2-} + 10\,H^+ + 10\,e^- \longrightarrow CH_4 + 3\,H_2O$$

Carbonate ion   Methane

$$NO_3^- + 2\,H^+ + 2\,e^- \longrightarrow NO_2^- + H_2O \xrightarrow{6\,H^+ + 6\,e^-} N_2O \xrightarrow{2\,H^+ + 2\,e^-} N_2$$

Nitrate ion   Nitrite ion   Nitrous oxide   Nitrogen gas

Figure 17.1
Chemical equations showing the reduction of some electron acceptors in bacterial respiration.

Plate streaking, Exercise 11

Carbohydrate catabolism, Exercise 13

Procedure

Nitrate Reduction Test

1. Label three tubes of nitrate broth, and inoculate one with *Lactococcus*, one with *Pseudomonas*, and one with *Bacillus*.
2. Incubate the tubes at 35°C for 2 days.
3. Add 5 drops of nitrate A and 5 drops of nitrate B to each tube and to an uninoculated nitrate broth tube. Shake the tubes gently.
4. A red color within 30 seconds is a positive test. If the broth turns red, what compound is present? ___
   _____
5. If it does not turn red, add a small pinch of zinc dust; if it turns red now, the test is negative. If not, it is positive for nitrate reduction. Why? _____
   _____
6. Record your results.

Oxidase Test

1. Divide one plate in half. Label one half *"Escherichia"* and the other *"Pseudomonas."* Streak each organism on the appropriate half.
2. Incubate the plate, inverted, at 35°C for 24 to 48 hours.

3. To test for cytochrome oxidase, do one of the following:
   a. Drop oxidase reagent on the colonies and observe for a color change to pink within 1 minute, then blue to black. Oxidase-negative colonies will not change color.
   b. Moisten an oxidase disk and place it *on* colonies. Incubate the plate for 15 minutes at 35°C. The colonies of oxidase-positive bacteria will turn black. (See color plate III.11.)

Catalase Test

1. Divide the other plate in half, and inoculate one half with *Lactococcus* and the other half with one short streak or spot of *Bacillus*.
2. Incubate the plate, inverted, at 35°C for 24 hours. After growth occurs, the plate may be kept at room temperature until the next lab period.
3. To test for catalase, do one of the following:
   a. Drop hydrogen peroxide on the colonies and observe for bubbles.
   b. With a sterile toothpick, touch the center of a colony and transfer it to a clean glass slide. Discard the toothpick in the To Be Autoclaved area. Drop hydrogen peroxide on the organisms and observe for bubbles (catalase-positive). (See color plate XII.3.) What gas is in the bubbles? _____
   _____

exercise 17

Respiration

Name _____

Date _____

Lab Section _____

Purpose _____

_____

_____

Data

Nitrate Reduction Test

Organism	Color After Adding Nitrate Reagents A and B	Color After Adding Zinc	Gas	NO$_3$ Reduction?
Uninoculated tube				
Lactococcus lactis				
Pseudomonas aeruginosa				
Bacillus subtilis				

Oxidase Test

Organism	Color After Adding Reagent	Oxidase Reaction
Escherichia coli		
Pseudomonas aeruginosa		

Catalase Test

Organism	Appearance After Adding H$_2$O$_2$	Catalase Reaction
Lactococcus lactis		
Bacillus subtilis		

Questions

1. Define reduction. _____

_____

2. Why does hydrogen peroxide bubble when it is poured on a skin cut? _____

_____

_____

3. Differentiate between aerobic respiration and anaerobic respiration. _____

_____

_____

4. Differentiate between fermentation and anaerobic respiration. _____

_____

_____

Critical Thinking

1. Would nitrate reduction occur more often in the presence or absence of molecular oxygen? Explain.

2. In the nitrate reduction test, what does the presence of gas indicate?

3. Is nitrate reduction beneficial or harmful to farmers?

exercise 18

Rapid Identification Methods

Objectives

After completing this exercise, you should be able to:

1. Evaluate three methods of identifying enterics.
2. Name three advantages of the "systems approach" over conventional tube methods.

Background

The clinical microbiology laboratory must identify bacteria quickly and accurately. Accuracy is improved by using a series of standardized tests. The IMViC tests were developed as a means of separating members of the Enterobacteriaceae (enterics), particularly the coliforms,* using a standard combination of four tests. Each capital letter in **IMViC** represents a test; the *i* is added for easier pronunciation. The tests are as follows:

I for indole production from tryptophan

M for methyl red test for acid production from glucose

V for the Voges–Proskauer test for production of acetoin from glucose

C for the use of citrate as the sole carbon source. The Simmons citrate agar used in this exercise contains the indicator bromthymol blue. Citric acid will be the only source of carbon; therefore, only organisms capable of using citric acid as a source of carbon will grow. When the citric acid is metabolized, an excess of sodium and ammonium ions results, and the indicator turns from green to blue, indicating alkaline conditions. (See color plate III.10.)

Although variation among strains does exist, IMViC reactions for selected species of the Enterobacteriaceae are given in Table 18.1.

Rapid identification methods have been developed which provide a large number of results from one inoculation. Examples are Enterotube® II** and API 20E®*** for identifying oxidase-negative, gram-negative bacteria belonging to the family Enterobacteriaceae. Enterotube® II (see color plate III.7) is divided into

*Enterobacteriaceae are aerobic or facultatively anaerobic, gram-negative, nonendospore-forming, rod-shaped bacteria. Coliforms are Enterobacteriaceae that ferment lactose with acid and gas formation within 48 hours at 35°C.

**BD Bioscience, Cockeysville, MD 21030.

***bioMérieux Vitek, Inc., Hazelwood, MO 63042.

Table 18.1

IMViC Reactions for Selected Species of the Enterobacteriaceae

Species	Indole	Methyl Red	Voges–Pros-kauer	Citrate
Escherichia coli	+(v)	+	–	–
Citrobacter freundii	–	+	–	+
Enterobacter aerogenes	–	–	+	+
Enterobacter cloacae	–	–	+	+
Serratia marcescens	–	+ or –*	+	+
Proteus vulgaris	+	+	–	–(v)
Proteus mirabilis	–	+	– or +**	+(v)

v = variable
* Majority of strains give + results
** Majority of strains give – results

twelve compartments, each containing a different substrate in agar. API 20E® (see color plate III.5) consists of twenty microtubes containing dehydrated substrates. The substrates are rehydrated by adding a bacterial suspension. No culturing beyond the initial isolation is necessary with these systems. Comparisons between these rapid identification methods and conventional culture methods show that they are as accurate as conventional test tube methods.

Computerized analysis of test results increases accuracy because each test is given a point value. Tests that are more important than others get more points. The IMViC tests are four tests of equal value.

As commercial identification systems are developed, they can provide greater standardization in identification because they overcome the limitations of hunting through a key, differences in media preparation, and evaluation of tests within a laboratory or between different laboratories. They also save time, money, and labor.

Materials

Petri plate containing nutrient agar

IMViC tests and reagents

Enterotube® II and reagents

API 20E® tray and reagents

Tube containing 5.0 ml sterile saline

5 ml pipette

Oxidase reagent

Mineral oil

Sterile Pasteur pipette

Culture

Unknown enteric # _____

Techniques Required

Inoculating loop and needle, Exercise 10

Aseptic technique, Exercise 10

MRVP tests, Exercise 14

Fermentation tests, Exercise 14

Protein catabolism, parts 1 and 2, Exercises 15 and 16

Respiration, Exercise 17

Procedure

Isolation

1. Streak the nutrient agar plate with your unknown for isolation and to determine purity of the culture. Incubate the plate, inverted, at 35°C for 24 to 48 hours. Record the appearance of the colonies.
2. Determine the oxidase reaction of one of the colonies remaining on the plate. Why? _____
   _____

   How will you determine the oxidase reaction? ___
   _____
   _____

IMViC Tests

1. Inoculate tubes of tryptone broth (indole test), MRVP broths, and Simmons citrate agar with your unknown.
2. Incubate the tubes at 35°C for 48 hours or longer; perform the appropriate tests, and record your results.

Enterotube® II (Figure 18.1)

1. Remove both caps from the Enterotube® II. One end of the wire is straight and is used to pick up the inoculum; the bent end of the wire is the handle.

(a) Pick a well-isolated colony with the inoculating end of the wire.

(b) Hold the bent end of the wire, and withdraw the needle through all twelve compartments with a turning motion.

(c) Reinsert the wire through all twelve compartments. Then withdraw it to the notch on the wire. Break the wire at the notch.

(d) Using the broken wire, punch holes through the foil covering the air inlets in the last eight compartments. Replace the caps loosely.

(e) After incubation, compare the tube to an uninoculated one to record results.

Figure 18.1
Inoculating an Enterotube® II.

Table 18.2
Enterotube® II Biochemical Reactions

Test	Comments	Indicator Changed	
		From	To
GLU	Acid from glucose	Red	Yellow
GAS	Gas produced from fermentation of glucose trapped in this compartment, causing separation of the wax		
LYS	Lysine decarboxylase	Yellow	Purple
ORN	Ornithine decarboxylase	Yellow	Purple
H_2S	Ferrous ion reacts with sulfide ions, forming a black precipitate		
IND	Kovacs reagent is added to the H_2S/IND compartment to detect indole	Beige	Red
ADON	Adonitol fermentation	Red	Yellow
LAC	Lactose fermentation	Red	Yellow
ARAB	Arabinose fermentation	Red	Yellow
SORB	Sorbitol fermentation	Red	Yellow
V–P	Voges–Proskauer reagents detect acetoin	Beige	Red
DUL	Dulcitol fermentation	Green	Yellow
PA	Phenylpyruvic acid released from phenylalanine after its deamination combines with iron salts to form a black precipitate		
UREA	Ammonia changes the pH of the medium	Yellow	Pink
CIT	Citric acid used as a carbon source	Green	Blue

Source: BD Bioscience, Cockeysville, MD 21030.

Holding the Enterotube® II, pick a well-isolated colony with the inoculating end of the wire (Figure 18.1a). Avoid touching the agar with the needle.

2. Inoculate the Enterotube® II by holding the bent end of the wire and twisting; the tip of the wire should be visible in the citrate compartment. Withdraw the needle through all twelve compartments using a turning motion (Figure 18.1b).

3. Reinsert the needle into the Enterotube® II, using a turning motion, through all twelve compartments until the notch on the wire is aligned with the opening of the tube. Break the wire at the notch by bending it (Figure 18.1c). The portion of the needle remaining in the tube maintains anaerobic conditions necessary for fermentation, production of gas, and decarboxylation.

4. Punch holes with the broken-off wire through the foil covering the air inlets of the last eight compartments (adonitol through citrate) to provide aerobic conditions (Figure 18.1d). Replace the caps on both ends of the tube.

Discard the handle in disinfectant.

5. Incubate the tube lying on its flat surface at 35°C for 24 hours.

6. Interpret and record all reactions (see Table 18.2) in the Laboratory Report. Read all the other tests *before* the indole and V–P tests, which follow.

   a. Indole test. Place the Enterotube® II horizontally and melt a small hole in the plastic film covering the H_2S/indole compartment using a warm inoculating loop. Add 1 to 2 drops of Kovacs reagent, and allow the reagent to contact the agar surface. A positive test is indicated by a red color within 10 seconds.

b.   V–P test. Add 2 drops of 20% KOH containing 5% α-naphthol to the V–P compartment. A positive test is indicated by development of a red color within 20 minutes.

7.   Indicate each positive reaction by circling the number appearing below the appropriate compartment of the Enterotube® II outlined in the Laboratory Report. Add the circled numbers only within each bracketed section, and enter this sum in the space provided below the arrow. Note that the V–P test is used as a confirming test only. Read the five numbers in these spaces across as a five-digit number in the *Computer Coding and Identification System*.*

8.   Dispose of the Enterotube® II by placing it in the autoclave bag.

API 20E® (Figure 18.2)

1.   Prepare a bacterial suspension by touching the center of a well-isolated colony with a sterile loop, and thoroughly mix the inoculum in 5 ml of sterile saline (Figure 18.2a).

2.   Place 5 ml of tap water into the corrugated incubation tray, to provide a humid atmosphere during incubation.

3.   Using a sterile Pasteur pipette, tilt the API 20E® tray and fill the tube section of the microtubes with the bacterial suspension. Fill the tube *and* cupule sections of the CIT, VP, and GEL tubes (Figure 18.2b).

4.   After inoculation, completely fill the cupule section of the ADH, LDC, ODC, H_2S, and URE tubes with mineral oil to create anaerobic conditions (Figure 18.2c).

5.   Place the plastic lid on the tray and incubate the strip for 24 hours at 35°C (Figure 18.2d). If the

*BD Bioscience, Cockeysville, MD 21030.

strip cannot be read after 24 hours, remove the strip from the incubator and refrigerate it.

6.   Interpret and record all reactions (see Table 18.3) in the Laboratory Report. Read all the other tests before the TDA, VP, and IND, which follow.

a.   TDA test. Add 1 drop 10% ferric chloride. A positive test is brownish-red. Indole-positive organisms may produce an orange color; this is a negative TDA reaction.

b.   V–P test. Add 1 drop of V–P reagent II (KOH), then 1 drop of V–P reagent I (α-napthol). A positive reaction produces a red color (not pale pink) after 10 minutes.

c.   Indole test. Add 1 drop of Kovacs reagent. A red ring after 2 minutes indicates a positive reaction.

d.   Nitrate reduction. Before adding reagents, look for bubbles in the GLU tube. Bubbles indicate reduction of nitrate to N_2. Add 2 drops of nitrate reagent A (dimethyl-α-napthylamine) and 2 drops of nitrate reagent B (sulfanilic acid). A positive reaction (red) may take 2 to 3 minutes to develop. A negative test can be confirmed by adding zinc dust.

7.   Indicate each positive reaction with a (+) in the appropriate compartment of the Laboratory Report. Add the points for each positive reaction within each bold-outlined section. Read the seven numbers across as a seven-digit number in the *API 20E Analytical Profile Index*** or use the automated Voice Response System.*** Nitrate reduction is a confirming test and not part of the seven-digit code.

8.   Dispose of the API strip, tray, and lid by placing them in the autoclave bag.

**bioMérieux Vitek, Inc., Hazelwood, MO 63042.

***The Voice Response System is available at 1-800-645-7056.

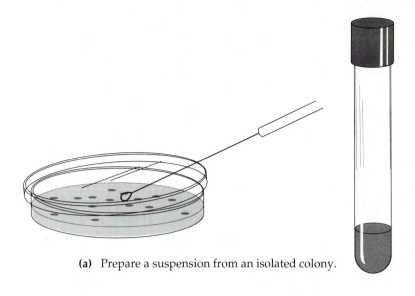

(a) Prepare a suspension from an isolated colony.

Dehydrated substrates

Cupule
Microtube

(b) Aseptically inoculate the microtubes with the bacterial suspension. Fill the tube and cupule for CIT, VP, and GEL (boxed tests).

(c) Add mineral oil to the ADH, LDC, ODC, H₂S, and URE cupules (underlined tests).

(d) Incubate the strip in its plastic tray.

Figure 18.2
Inoculating the API 20E® system.

Table 18.3

API 20E® Biochemical Reactions

Test	Comments	Indicator	
		Positive	Negative
ONPG	O-nitrophenyl-β-D-galactopyranoside is hydrolyzed by the enzyme that hydrolyzes lactose	Yellow	Colorless
ADH	Arginine dihydrolase transforms arginine into ornithine, NH_3, and CO_2	Red	Yellow
LDC	Decarboxylation of lysine liberates cadaverine	Red	Yellow
ODC	Decarboxylation of ornithine produces putrescine	Red	Yellow
CIT	Citric acid used as sole carbon source	Dark blue	Light green
H_2S	Blackening indicates reduction of thiosulfate to H_2S	Black	No blackening
URE	Urea is hydrolyzed by the enzyme urease to NH_3 and CO_2	Red	Yellow
TDA	Deamination of tryptophan produces indole and pyruvic acid	Brown	Yellow
IND	Kovacs reagent is added to detect indole	Red ring	Yellow
VP	Addition of KOH and α-naphthol detects the presence of acetoin	Red	Colorless
GEL	Gelatin hydrolysis	Diffusion of pigment	No diffusion
GLU	Fermentation of glucose		
MAN	Fermentation of mannitol		
INO	Fermentation of inositol		
SOR	Fermentation of sorbitol		
RHA	Fermentation of rhamnose	Yellow or yellow-green	Blue or green
SAC	Fermentation of sucrose		
MEL	Fermentation of melibiose		
AMY	Fermentation of amygdalin		
ARA	Fermentation of arabinose		
NO_2 N_2 gas	Nitrate reduction	Red Bubbles Yellow after addition of zinc	Yellow No bubbles Orange after reagents and zinc

Source: bioMérieux Vitek Inc., Hazelwood, MO 63042.

exercise 18

Rapid Identification Methods

Name _____

Date _____

Lab Section _____

Purpose _____

_____

_____

Data

Unknown # _____

Appearance on nutrient agar: _____

Oxidase reaction: _____

IMViC

Indicate positive (+) and negative (−) results for each test.

Indole: _____

Methyl red: _____

V–P: _____

Citrate: _____

Enterotube® II

Circle the number corresponding to each positive reaction below the appropriate compartment. Then determine the five-digit code.

ENTEROTUBE® II

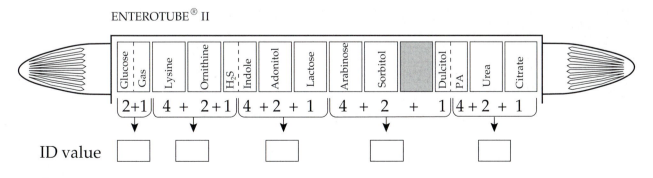

ID value

V–P: _____

API 20E®

Indicate positive (+) and negative (−) results in the Results line. Then determine the seven-digit code.

	ONPG 1	ADH 2	LDC 4	ODC 1	CIT 2	H₂S 4	URE 1	TDA 2	IND 4	VP 1	GEL 2	GLU 4
Results												
Profile number												

	MAN 1	INO 2	SOR 4	RHA 1	SAC 2	MEL 4	AMY 1	ARA 2	Oxidase 4	NO₂	N₂ gas
Results											
Profile number											

Questions

1. What species was identified in unknown # _____ by the IMViC tests? _____

    By the API 20E®? _____

    By the Enterotube® II? _____

2. Did all three methods agree? _____ If not, explain any discrepancies. _____

    _____

    _____

3. Which method did you prefer? _____ Why? _____

    _____

    _____

4. Why are systems developed to identify Enterobacteriaceae? _____

    _____

    _____

5. Why is an oxidase test performed on a culture before using API 20E® and Enterotube® II to identify the culture?

    _____

    _____

Critical Thinking

1. Why should the first digit in the five-digit Enterotube® II ID value be equal to or greater than 2, and the fourth digit in the API 20E® profile number be equal to or greater than 4?

2. Why can one species have two or more numbers in a rapid identification system? (For example, *E. coli* is Enterotube® II numbers 34540 and 34560; *Citrobacter braakii* is API 20E® numbers 3504552 and 3504553.)

3. Use Table 18.1 to give an example of a limitation of the IMViC tests.

Microbial Growth

EXERCISES

In previous exercises, we studied bacterial growth by observing colonies or turbidity, making visual determinations on an all-or-none basis with the naked eye. In Exercise 20, we will use instrumentation to detect the slight changes in bacterial numbers that occur over time—changes that are not visible to the naked eye—to measure lag and log phases of growth (see the graph on the next page).

Bacteria have nutritional, physical, chemical, and environmental requirements that must be met in order for growth to occur. A knowledge of the conditions necessary for microbial growth can facilitate culturing microorganisms and controlling unwanted organisms.

In 1861, Jean Baptiste Dumas presented a paper to the Academy of Sciences on behalf of Pasteur. In this paper, he stated:

> *The existence of infusoria having the characteristics of ferments is already a fact which seems to deserve attention, but a characteristic which is even more interesting is that these infusoria-animalcules live and multiply indefinitely in absence of the smallest amount of air or free oxygen. . . . This is, I believe, the first known example of animal ferments and also of animals living without free oxygen gas.**

---

*In H. A. Lechevalier and M. Solotorovsky. *Three Centuries of Microbiology*. New York: Dover Publications, 1974, p. 45.

When we speak of *air* as a growth requirement, we are referring to the oxygen in air. Furthermore, when we refer to oxygen, we usually mean the **molecular oxygen (O_2)** that acts as an electron acceptor in aerobic respiration. The significance of the presence or absence of oxygen is investigated in Exercise 19.

Physical environmental conditions that influence microbial growth are temperature (Exercise 20), osmotic pressure (Exercise 21), and pH (Exercise 21).

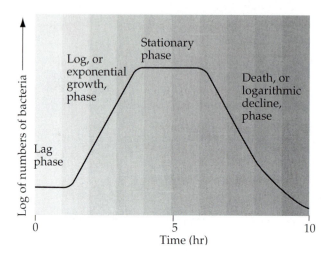

Bacterial growth curve, showing the four typical phases of growth.

exercise 19

Oxygen and the Growth of Bacteria

What is education but a process by which a person begins to learn how to learn?

PETER USTINOV

Objectives

After completing this exercise, you should be able to:

1. Identify the incubation conditions for each of the following types of organisms: aerobes, obligate anaerobes, aerotolerant anaerobes, microaerophiles, and facultative anaerobes.
2. Describe three methods of culturing anaerobes.
3. Cultivate anaerobic bacteria.

Background

The presence or absence of molecular oxygen (O_2) can be very important to the growth of bacteria. Some bacteria, called obligate **aerobic bacteria,** require oxygen, while others, called **anaerobic bacteria,** do not use oxygen. One reason **obligate anaerobes** cannot tolerate the presence of oxygen is that they lack catalase, and the resultant accumulation of hydrogen peroxide

is lethal. **Aerotolerant anaerobes** cannot use oxygen but tolerate it fairly well, although their growth may be enhanced by microaerophilic conditions. Most of these bacteria use fermentative metabolism.

Some bacteria, the **microaerophiles,** grow best in an atmosphere with increased carbon dioxide (7% to 10%) and lower concentrations of oxygen. Microaerophiles will grow in a solid nutrient medium at a depth to which small amounts of oxygen have diffused into the medium (Figure 19.1e). In order to culture microaerophiles on Petri plates and nonreducing media, a **candle jar** is used. Inoculated plates and tubes are placed in a large jar with a lighted candle. After the lid is placed on the jar, the candle will extinguish when the concentration of oxygen decreases (the concentration of carbon dioxide will be raised).

The majority of bacteria are capable of living with or without oxygen; these bacteria are called **facultative anaerobes** (Figure 19.1b).

(a) Obligate aerobes (b) Facultative anaerobes (c) Obligate anaerobes (d) Aerotolerant anaerobes (e) Microaerophiles

Figure 19.1

Effect of oxygen concentration on the growth of various types of bacteria in a tube of solid medium. Oxygen diffuses only a limited distance from the atmosphere into the solid medium.

Five genera of bacteria lacking catalase are *Strepto-coccus, Enterococcus, Leuconostoc, Lactobacillus,* and *Clostridium.* Species of *Clostridium* are obligate anaerobes, but members of the other four genera are aerotolerant anaerobes. The four genera of aerotolerant anaerobes lack the cytochrome system to produce hydrogen peroxide and therefore do not need catalase. Determining the presence or absence of catalase can help in identifying bacteria. When a few drops of 3% hydrogen peroxide are added to a microbial colony and catalase is present, molecular oxygen is released as bubbles.

$$2 H_2O_2 \longrightarrow 2 H_2O + O_2$$

In the laboratory, we can culture anaerobes either by excluding free oxygen from the environment or by using reducing media. Many anaerobic culture methods involve both processes. Some of the anaerobic culturing methods are as follows:

1. **Reducing media** contain reagents that chemically combine with free oxygen, reducing the concentration of oxygen. In thioglycollate broth, **sodium thioglycollate** ($HSCH_2COONa$) will combine with oxygen. A small amount of agar is added to increase the viscosity, which reduces the diffusion of air into the medium. Usually dye is added to indicate where oxygen is present in the medium. Resazurin, which is pink in the presence of excess oxygen and colorless when reduced, or methylene blue (see below) are commonly used indicators.

2. Conventional, nonreducing media can be incubated in an anaerobic environment. Oxygen is excluded from a **Brewer anaerobic jar** by adding a Gas-Pak® and a palladium catalyst to catalytically combine the hydrogen with oxygen and form water (Figure 19.2). Carbon dioxide and hydrogen are given off when water is added to the Gas-Pak® envelope of sodium bicarbonate and sodium borohydride. A methylene blue indicator strip is placed in the jar; methylene blue is blue in the presence of oxygen and white when reduced. One of the disadvantages of the Brewer jar is that the jar must be opened to observe or use one plate. An inexpensive modification of the Brewer jar has been developed using disposable plastic bags for one or two culture plates (Figure 19.3). In this technique, the bag is coated with antifogging chemicals, and a wet sodium bicarbonate tablet is added to generate carbon dioxide. Iron (Fe^0) activated with water removes O_2 as iron oxide (Fe_2O_3) is produced. One plate can be observed without opening the bag.

3. **Anaerobic incubators** and **glove boxes** can also be used for incubation. Air is evacuated from the chamber and can be replaced with a mixture of carbon dioxide and nitrogen.

Figure 19.2
Brewer anaerobic jar. Carbon dioxide and hydrogen are generated by the addition of water to a chemical packet. Hydrogen gas combines with atmospheric oxygen in the presence of a palladium catalyst (in the lid or in the Gas-Pak® Plus packet) to form H_2O. The anaerobic indicator contains methylene blue, which is blue when oxidized and turns colorless when the oxygen is removed from the container.

Figure 19.3
Plastic bag with Petri plate and CO_2 generator.

All methods of anaerobic culturing are only effective if the specimen or culture of anaerobic organisms is collected and transferred in a manner that minimizes exposure to oxygen. In this exercise, we will try two methods of anaerobic culture and will perform the catalase test.

Materials

Petri plates containing nutrient agar (2)

Tubes containing thioglycollate broth with indicator (4)

Brewer anaerobic jars (2 per lab section)

3% hydrogen peroxide (H_2O_2) (second period)

Cultures

Alcaligenes faecalis
Clostridium sporogenes
Enterococcus faecalis
Escherichia coli

Techniques Required

Inoculating loop, Exercise 10

Aseptic technique, Exercise 10

Plate streaking, Exercise 11

Selective media, Exercise 12

Catalase test, Exercise 17

Procedure

1. Don't shake the thioglycollate. Why? _____

   _____

   What should you do to salvage it if you do shake it? _____

   _____

2. Label four tubes of thioglycollate broth, and aseptically inoculate one with a loopful of *Alcaligenes*, one with *Clostridium*, one with *Enterococcus*, and one with *Escherichia*.

3. Incubate the tubes at 35°C until the next lab period.

4. Record the appearance of growth in each tube.

5. With a marker, divide each nutrient agar plate into four sectors on the bottom of the plate. Label one plate "Aerobic" and the other "Anaerobic."

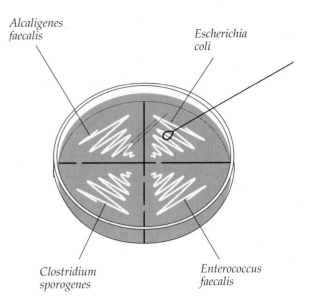

Alcaligenes faecalis

Escherichia coli

Clostridium sporogenes

Enterococcus faecalis

Figure 19.4
Divide a Petri plate into quadrants by drawing lines on the bottom of the plate. Inoculate each quadrant by streaking it with an inoculating loop.

6. Streak a single line of *Alcaligenes, Clostridium, Enterococcus*, and *Escherichia* (Figure 19.4).

7. Incubate the "Aerobic" plate, inverted, at 35°C until the next lab period. Place the "Anaerobic" plate, inverted, in the Brewer jar.

8. Your instructor will demonstrate how the jar is rendered anaerobic once it is filled with plates. Ten ml of water are added to a Gas-Pak® generator and placed in the jar with a methylene blue indicator. What causes the condensation that forms on the sides of the jar? _____

   _____

9. Incubate the jar at 35°C until the next lab period.

10. After incubation, record the growth on each plate.

11. Perform the catalase test by adding a few drops of 3% H_2O_2 to the different colonies on the nutrient agar. A positive catalase test produces a bubbling white froth. A dissecting microscope (Appendix E) can be used if more magnification is required to detect bubbling. The catalase test may also be done by transferring bacteria to a slide and adding the H_2O_2 to it. Bacterial growth on blood agar must be tested this way. Why? _____

    _____

exercise 19

LABORATORY REPORT

Oxygen and the Growth of Bacteria

Name _____

Date _____

Lab Section _____

Purpose _____

_____

_____

Expected Results

Before the next lab period, indicate the amount of growth you *expect* to see on each plate and the catalase reaction.

Species	Aerobic		Anaerobic	
	Growth	Catalase Reaction	Growth	Catalase Reaction
Alcaligenes				
Clostridium				
Enterococcus				
Escherichia				

Data

Thioglycollate

Sketch the location of bacterial growth and note where the indicator has turned blue or pink.

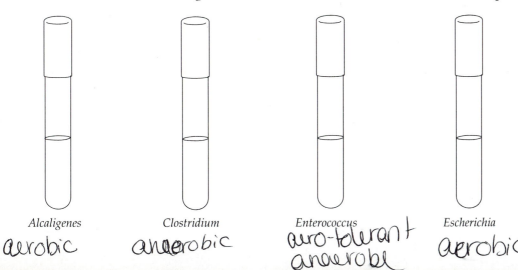

Alcaligenes
aerobic

Clostridium
anaerobic

Enterococcus
aero-tolerant anaerobe

Escherichia
aerobic

Nutrient Agar Plates

Species	Aerobic		Anaerobic	
	Growth	Catalase Reaction	Growth	Catalase Reaction
Alcaligenes				
Clostridium				
Enterococcus				
Escherichia				

Does the anaerobic plate have an odor? _____

Conclusions

Categorize the four organisms used in this exercise in terms of their oxygen requirements:

Alcaligenes: _____

Clostridium: _____

Enterococcus: _____

Escherichia: _____

Questions

1. Did your results differ from your expected results? _____

   Briefly explain why or why not. _____

   _____

2. Can aerobic bacteria grow in the absence of O_2? _____ What would you need to do to determine

   whether bacteria growing on a Petri plate from the Brewer jar are anaerobes? _____

3. What does the appearance of a blue or pink color in a thioglycollate tube mean? _____

   _____

   _____

4. Why will obligate anaerobes grow in thioglycollate? _____

   _____

   _____

5. To what do you attribute the odors of anaerobic decomposition? _____

_____

_____

Critical Thinking

1. *Bergey's Manual* describes *Streptococcus* and *Escherichia* as facultative anaerobes. How do the oxygen requirements of these organisms differ? Which one could correctly be called an aerotolerant anaerobe?

2. How can the aerobe *Pseudomonas aeruginosa* grow in the absence of oxygen?

3. The catalase test is often used clinically to distinguish between two genera of gram-positive cocci, _____

and _____ . It is also used to distinguish between two genera of gram-

positive rods, _____ and _____ .

4. The following genera of anaerobes are commonly found in fecal samples. Use the key below to differentiate them: *Bacteroides, Fusobacterium, Clostridium, Enterococcus, Veillonella,* and *Lactobacillus*.

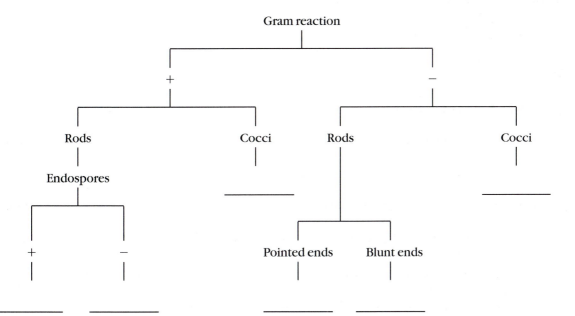

exercise 20

Determination of a Bacterial Growth Curve: The Role of Temperature

Objectives

After completing this exercise, you should be able to:

1. Identify the four phases of a typical bacterial growth curve.
2. Measure bacterial growth turbidometrically.
3. Interpret growth data plotted on a graph.
4. Determine the effect of temperature on bacterial growth.

Background

The phases of growth of a bacterial population can be determined by measuring the turbidity of a broth culture (see the figure in the introduction to Part Five). Turbidity is not a direct measure of bacterial numbers, but increasing turbidity does indicate bacterial growth. Since millions of cells per milliliter must be present to discern turbidity with the eye, a spectrophotometer is used to detect the presence of smaller numbers of bacteria (Appendix C).

Most bacteria grow within a particular temperature range (Figure 20.1a). The *minimum growth temperature* is the lowest temperature at which a species will grow. A species grows fastest at its *optimum growth temperature*. And the highest temperature at which a species can grow is its *maximum growth temperature*. Some heat is necessary for growth. Heat probably increases the rate of collisions between molecules, thereby increasing enzyme activity. At temperatures near the maximum growth temperature, growth ceases, presumably due to inactivation of enzymes.

Bacteria can be classified into three primary groups, based on their growth at various temperatures (Figure 20.1b). The optimum growth temperature for **psychrophiles** is about 15°C. The optimum growth temperature for **mesophiles** is between 25°C and 40°C. The optimum growth temperature for many **thermophiles** is between 45°C and 65°C, although some are capable of growth above 90°C. The range of temperature preferred by bacteria is genetically determined, resulting in enzymes with different temperature requirements.

In this exercise, we will plot growth curves for a bacterium at different temperatures to determine the preferred temperature range of this species.

(a)

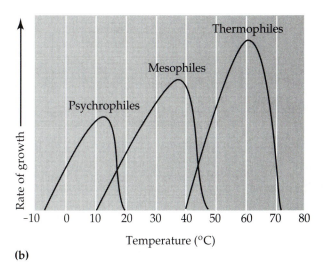

(b)

Figure 20.1
(a) Growth response of bacteria within their growing temperature range. (b) Typical growth responses to temperature for psychrophiles, mesophiles, and thermophiles.

Materials

Flask containing nutrient broth + 1.5% NaCl

Spectrophotometer tubes (2)

Sterile 5 ml pipettes (11)

Sterile 10 ml pipette

Spectrophotometer

Culture

Escherichia coli

Techniques Required

Inoculating loop, Exercise 10

Aseptic technique, Exercise 10

Pipetting, Appendix A

Spectrophotometry, Appendix C

Graphing, Appendix D

Procedure

1. Aseptically transfer 4 ml nutrient broth to one of the spectrophotometer tubes. This is your control to *standardize* the spectrophotometer.

2. Aseptically inoculate a flask of nutrient broth with 10 ml of *E. coli.*

3. Swirl to mix the contents. Transfer 4 ml from the flask to the second spectrophotometer tube, and measure the absorbance (Abs.) on the spectrophotometer. *Read* Appendix C before using the spectrophotometer.

4. Record all measurements.

> **After taking a reading, discard the 4 ml into disinfectant.**

5. Place the flask at your *assigned* temperature.

6. Record the absorbance every 10 minutes for 60 to 90 minutes. Remove the flask from the water bath for the minimum time possible to take each sample. To take a sample, aseptically pipette 4 ml into the second spectrophotometer tube.

7. Graph the data you obtained. *Read* Appendix D before drawing your graph.

exercise 20

Determination of a Bacterial Growth Curve: The Role of Temperature

Name _____

Date _____

Lab Section _____

Purpose _____

_____

_____

Data

Temperature: _____

Time (min)	Absorbance	Time (min)	Absorbance
0			

Plot your data on the graph paper. Record absorbance on the Y-axis and time on the X-axis. Obtain data for other temperatures from your classmates. Plot these data on the same graph using different lines or colors. Label the phases of growth.

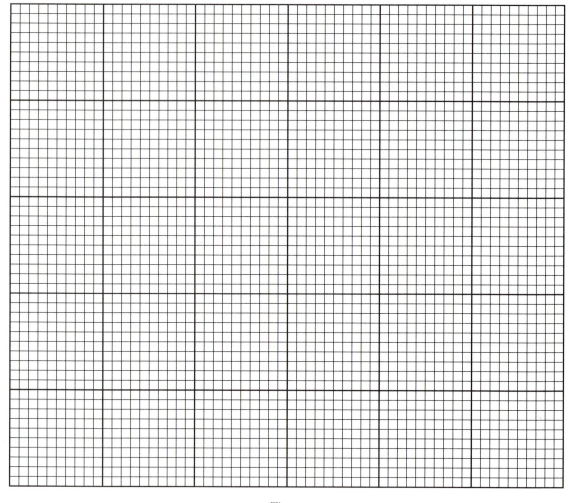

Absorbance

Time

Conclusions

1. Summarize the effect of temperature on growth of *E. coli.* _____

_____

_____

_____

2. From the class data, what is *E. coli*'s optimum temperature? _____ Does this agree with your textbook? _____

If not, provide a brief explanation. _____

Questions

1. Why aren't you likely to get lag and death phases in this experiment? _____

_____

_____

2. What is the optimum growth temperature for human pathogens? _____

3. Where in nature would you most likely find a thermophile? _____

A psychrophile? _____

4. What is the effect of temperature on enzymes? _____

_____

Can you use any examples from your experiments? _____

Critical Thinking

1. The following graph shows a likely relationship between bacterial growth and oxygen use. Explain the relationship.

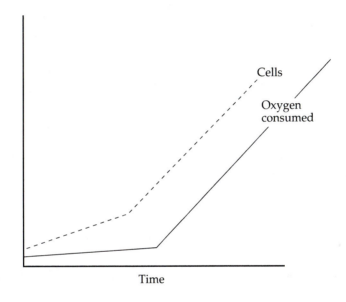

2. The following data were obtained for a log phase culture using plate counts and spectrophotometry:

No. of Cells/ml	Absorbance
5×10^4	0.04
5×10^5	0.26
6×10^7	0.72

Assume that you inoculated a flask yesterday and now you need a broth culture containing 5×10^6 cells/ml to perform an experiment. The concentration of cells in the 24-hour culture is much too high; its absorbance is greater than 1. What is the easiest way to get the desired cell concentration from your 24-hour flask?

3. You have isolated a bacterium that produces an antibiotic. The bacterium grows well at 37°C but doesn't produce the antibiotic at that temperature. Antibiotic production is best at 30°C, but the cells grow so poorly at that temperature that you can't get enough antibiotic. The relationships are shown in the following graphs. Design a procedure to maximize the biomass (total cells) and the yield of the antibiotic.

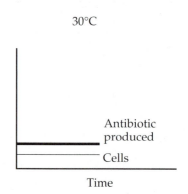

4. Why is turbidity not an accurate measurement of viable bacteria in a culture?

exercise 21

Other Influences on Microbial Growth: Osmotic Pressure and pH

EXERCISE
21

Objectives

After completing this exercise, you should be able to:

1. Define osmotic pressure and explain how it affects a cell.
2. Explain how microbial growth is related to osmotic pressure and pH.
3. Prepare and use a gradient plate.

Background

The osmotic pressure of an environment influences microbial growth. **Osmotic pressure** is the force with which a solvent (water) moves from a solution of lower solute concentration to a solution of higher solute concentration across a semipermeable membrane (Figure 21.1). The addition of solutes, especially salts and sugars, and the resultant increase in osmotic pressure, are used to preserve some foods. Can you name a food that is preserved with salt? _____
With sugar? _____

Bacteria are often able to live in a **hypotonic** (hypoosmotic) environment, in which the concentration of solutes outside the cell is lower than that inside the cell. However, when the concentration of solutes outside the cell is higher than that inside the cell—a **hypertonic** (hyperosmotic) environment—a bacterial cell will undergo plasmolysis. *Plasmolysis* occurs when water leaves the cell and the cytoplasmic membrane draws inward, away from the cell wall. A few bacteria, called **facultative halophiles,** are able to tolerate salt concentrations up to 10%, and **extreme halophiles** require 15% to 20% salt.

The acidity or alkalinity (**pH**) of the environment also influences microbial growth. Optimal bacterial growth usually occurs between pH 6.5 and pH 7.5. Only a few bacteria grow at an acidic pH below 4.0. Therefore, organic acids are used to preserve foods such as fermented dairy products.

Many bacteria produce acids that may inhibit their growth. **Buffers** are added to culture media to neutralize these acids. The peptones in complex media act as buffers. Phosphate salts are often used as buffers in chemically defined media.

In this exercise, we will compare bacterial growth to fungal growth. Fungi differ from bacteria morphologically and biochemically. The yeast and mold used here are microbes belonging to the Kingdom Fungi.

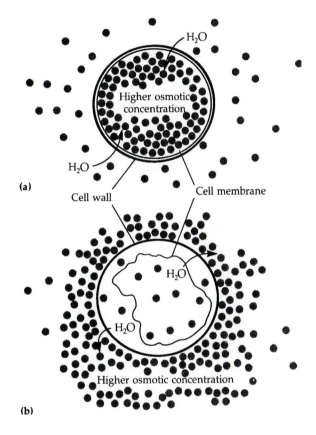

(a)

(b)

Figure 21.1
Osmotic pressure. **(a)** Water moves into the cell in a hypotonic solution. **(b)** Water leaves the cell in a hypertonic environment.

Materials

Effect of Osmotic Pressure on Growth

(Your instructor will assign salt, sugar, or nutrient agar control.)

Petri plates containing nutrient agar + 5%, 10%, 15% NaCl (one of each) *or*

Petri plates containing nutrient agar + 10%, 25%, 50% sucrose (one of each) *or*

Petri plate containing nutrient agar

Gradient Plate

Sterile Petri dish

Tube containing melted nutrient agar

Tube containing melted nutrient agar + 25% NaCl *or* 50% sucrose

Salt *or* sugar enrichments

Effect of pH on Growth

Tubes containing nutrient broth adjusted to pH 2.5, 5.0, 7.0, 9.5 (one of each)

Cultures

Effect of Osmotic Pressure on Growth

Escherichia coli

Staphylococcus aureus

A yeast (*Saccharomyces*)

A mold (*Penicillium* or *Aspergillus*)

Effect of pH on Growth (as assigned)

Staphylococcus aureus

Alcaligenes faecalis

Escherichia coli

Serratia marcescens

A mold (*Penicillium* or *Aspergillus*)

A yeast (*Saccharomyces*)

Techniques Required

Inoculating loop, Exercise 10

Plate streaking, Exercise 11

Procedure

Effect of Osmotic Pressure on Growth

1. Obtain a set of nutrient agar + salt *or* nutrient agar + sucrose plates. Draw two crossed lines to form four quadrants on the bottom of one plate. Each quadrant will be inoculated with one of the organisms and should be labeled accordingly (Figure 21.2). Repeat this procedure until all plates are marked. One student group will mark the *control* plate (nutrient agar alone). What is the purpose of the control plate? _____
   _____

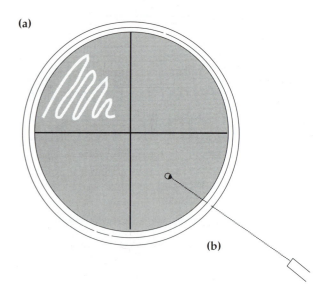

(a)

(b)

Figure 21.2
Inoculate each plate with four organisms. **(a)** Streak bacteria and yeast cultures. **(b)** Place a loopful of mold suspension in the center of one quadrant.

2. Using a sterile loop, inoculate *S. aureus* onto one quadrant on each plate (Figure 21.2a). Repeat this procedure to inoculate *E. coli* and the yeast.
3. Place a loopful of the mold suspension in the center of the remaining quadrant on each plate (Figure 21.2b).
4. Invert the plates and incubate them at 35°C for 24 to 48 hours. Record the relative amounts of bacterial or yeast growth; the culture showing the "best" growth is given (4+) and others are evaluated relative to that culture.
5. Incubate the plates at room temperature for 2 to 4 days, and record relative amounts of mold growth.

Gradient Plate

1. Pour a layer of nutrient agar into a sterile Petri dish and let it solidify with the dish resting on a small pencil or a loop handle (Figure 21.3a) so that a wedge forms.
2. On the bottom of the dish, draw a line at the end corresponding to the high side of the agar and label it "low." Draw four lines perpendicular to the first line, and if assigned sucrose, label the lines "0.5," "5," "10," and "15"; if assigned salt, "0.5," "5," "15," and "30" (Figure 21.3b).
3. Pour a layer of salt *or* sugar agar over the nutrient agar wedge. Let it solidify. A solute gradient concentration has been prepared. Why was the high side of the agar labeled "low" in step 2? _____
   _____

(a) Allow the nutrient agar to solidify into a wedge.

(b) Mark and label the plate.

Figure 21.3
Gradient plate.

4. The following enrichments have been prepared:
   a. Raisins were placed in tubes of nutrient broth + 0.5% sucrose, 5% sucrose, 10% sucrose, and 15% sucrose. The enrichments were incubated for 1 week.
   b. Hamburger was inoculated into tubes of nutrient broth + 0.5% NaCl, 5% NaCl, 15% NaCl, and 30% NaCl. The enrichments were incubated for 1 week.
5. Using the enrichment with the same solute as your gradient plate, streak a loopful of each enrichment onto the agar over the prelabeled lines. Streak from the low to the high end.
6. Incubate the plate, inverted, at 35°C for 48 hours, and record the growth patterns.

Effect of pH on Growth

1. Inoculate a loopful of your organism into each tube of pH test broth.
2. Incubate the tubes at 35°C for 24 to 48 hours. Record your results and results for the organisms tested by other students.

exercise 21

Other Influences on Microbial Growth: Osmotic Pressure and pH

Name _____

Date _____

Lab Section _____

Purpose _____

_____

_____

Data

Effect of Osmotic Pressure on Growth

Rate the relative amounts of growth on nutrient agar and nutrient agar containing solutes (salt and sucrose): $(-)$ = no growth; $(+)$ = minimal growth; $(2+)$ = moderate growth; $(3+)$ = heavy growth; $(4+)$ = very heavy (maximum) growth.

Medium	Organism/Amount of Growth			
	S. aureus	*E. coli*	Yeast	Mold
Nutrient agar				
+ 5% NaCl				
+ 10% NaCl				
+ 15% NaCl				
+ 10% sucrose				
+ 25% sucrose				
+ 50% sucrose				

Gradient Plate

Sketch your growth patterns and those of another group.

Low

Salt

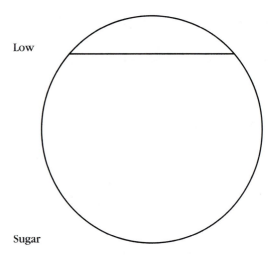

Low

Sugar

Effect of pH on Growth

Record the growth (+) or lack of growth (−) of each organism used.

pH	S. aureus	A. faecalis	E. coli	S. marcescens	Mold	Yeast
			Growth of			
2.5	−	−	−		−	
5.0	−	−	−		−	
7.0	+	+	+		−	
9.5	−	+	+		+	

Conclusions

1. Which organism tolerates high concentrations of salt best? _____

    Of what advantage is this to the organism? _____

    _____

2. Which organism tolerates high concentrations of sucrose best? _____

    Why is this advantageous to the organism? _____

    _____

3. Write a statement comparing the osmotic pressure tolerance of bacteria to that of fungi. _____

_____

4. What is the pH tolerance of bacteria compared to that of fungi? _____

_____

Questions

1. Explain the growth patterns on the gradient plates. _____

_____

_____

_____

_____

2. What is the principle of the gradient plate? _____

_____

3. What is the purpose of the nutrient agar control plate? _____

_____

4. Should (4+) growth occur only on the control plate? _____

Briefly explain. _____

_____

5. Why were the media containing raisins and hamburger labeled "enrichments"? _____

_____

6. Why are bacteria more likely to survive in a hypotonic environment than in a hypertonic environment? _____

_____

7. What is meant by "neutral pH"? _____

_____

8. Did any organisms grow in acidic conditions? _____ Why would you expect some

organisms to grow in an acidic environment? _____

_____

_____

_____

Critical Thinking

1. You are growing a bacterial culture in order to harvest a particular enzyme that the bacteria make. However, the bacteria make enough lactic acid within a few hours to kill themselves. What can you do to keep the cells growing?

2. *Helicobacter pylori* grows in the human stomach. The bacterium produces large quantities of urease. Of what value is this enzyme to the bacterium? What disease does *Helicobacter* cause?

3. The graph shows the pH (dashed line) and absorbance (solid line) of a bacterial culture over a 24-hour period. The culture medium consists of glucose and peptone. Explain the change in pH.

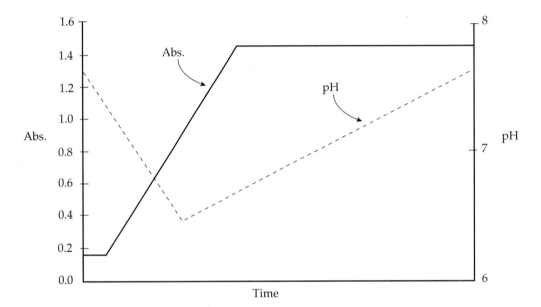

part six

Control of Microbial Growth

EXERCISES

The destruction of microbes by heat was employed by Lazzaro Spallanzani in the 1760s. He heated nutrient broth to kill preexisting life in his attempts to disprove the concept of spontaneous generation. In the 1860s, Pasteur heated broth in specially designed flasks and ended the debate over spontaneous generation (see the illustration on the next page). In Exercise 22, we will examine the effectiveness of heat for killing microbes.

The surgeon Joseph Lister was greatly influenced by Pasteur's demonstrations of the omnipresence of microorganisms and his proof that microorganisms cause decomposition of organic matter. Lister had observed the disastrous consequences of compound bone fractures (in which the skin is broken) compared to the relative safety of simple bone fractures. He had heard of the treatment of sewage with carbolic acid (phenol) to prevent diseases in villages that had sewage spills. In 1867, Lister wrote: "It appears that all that is requisite is to dress the wound with some material capable of killing these septic germs"* to prevent disease and death due to microbial growth.

*Quoted in W. Bulloch. *The History of Bacteriology.* New York: Dover Publications, 1974, p. 46.

In Part Six, we will examine current methods of controlling microbial growth, with special attention focused on the use of heat (Exercise 22), ultraviolet radiation (Exercise 23), disinfectants (Exercise 24), antibiotics (Exercise 25), and hand scrubbing (Exercise 26).

Pasteur's experiment that disproved spontaneous generation involved boiling nutrient broth in an S-neck flask. Microorganisms did not appear in the cooled broth even after a long period of time.

exercise 22

Physical Methods of Control: Heat

The successful man lengthens his stride when he discovers that the sign post has deceived him; the failure looks for a place to sit down.

JOHN R. ROGERS

Objectives

After completing this exercise, you should be able to:

1. Compare the bactericidal effectiveness of dry heat and moist heat on different species of bacteria.
2. Evaluate the heat tolerance of microbes.
3. Define and provide a use for each of the following: incineration, hot air oven, pasteurization, boiling, and autoclaving.

Background

The use of extreme temperature to control the growth of microbes is widely employed. Generally, if heat is applied, bacteria are killed; if cold temperatures are used, bacterial growth is inhibited.

Bacteria exhibit different tolerances to the application of heat. Heat sensitivity is genetically determined and is partially reflected in the optimal growth ranges (Exercise 20), which are **psychrophilic** (0°C to 20°C), **mesophilic** (25°C to 40°C), and **thermophilic** (45°C to 65°C), and by the presence of heat-resistant endospores. Overall, bacteria are more heat resistant than most other forms of life. Heat sensitivity of organisms can be affected by container size, cell density, moisture content, pH, and medium composition.

Heat can be applied as dry or moist heat. **Dry heat,** such as that in hot air ovens or incineration (for example, flaming loops), denatures enzymes, dehydrates microbes, and kills by oxidation effects. A standard application of dry heat in a hot air oven is 170°C for 2 hours. The heat of hot air is not readily transferred to a cooler body such as a microbial cell. Moisture transfers heat energy to the microbial cell more efficiently than dry air, resulting in the denaturation of enzymes. **Moist heat** methods include pasteurization, boiling, and autoclaving. In **pasteurization** the temperature is maintained at 63°C for 30 minutes or 72°C for 15 seconds to kill designated organisms that are pathogenic or cause spoilage. **Boiling** (100°C) for 10 minutes will kill vegetative bacterial cells; however, endospores are not in-

Table 22.1

Relationship Between Pressure and Temperature of Steam

Pressure (pounds per square inch, psi, in excess of atmospheric pressure)	Temperature (°C)
0 psi	100°C
5 psi	110°C
10 psi	116°C
15 psi	121°C
20 psi	126°C
30 psi	135°C

Source: G. J. Tortora, B. R. Funke, and C. L. Case. *Microbiology: An Introduction,* 7th ed. San Francisco, CA: Benjamin Cummings, 2001.

activated. The most effective method of moist heat sterilization is **autoclaving,** the use of steam under pressure. Increased pressure raises the boiling point of water and produces steam with a higher temperature (Table 22.1). Standard conditions for autoclaving are 15 psi, 121°C for 15 minutes. This is usually sufficient to kill endospores and render materials sterile.

There are two different methods of measuring heat effectiveness. **Thermal death time (TDT)** is the length of time required to kill all bacteria in a liquid culture at a given temperature. The less common **thermal death point (TDP)** is the temperature required to kill all bacteria in a liquid culture in 10 minutes.

Materials

Petri plates containing nutrient agar (2)

Thermometer

Empty tube

Beaker

Hot plate or tripod and asbestos pad

Ice

Cultures (as assigned)

Group A:
    Old (48 to 72 hours) *Bacillus subtilis*
    Young (24 hours) *Bacillus subtilis*

Group B:
    Staphylococcus epidermidis
    Escherichia coli

Group C:
    Young (24 hours) *Bacillus subtilis*
    Escherichia coli

Group D:
    Mold (*Penicillium*) spore suspension
    Old (48 to 72 hours) *Bacillus subtilis*

Demonstration

Autoclaved and dry-heated soil

Techniques Required

Inoculating loop, Exercise 10

Aseptic technique, Exercise 10

Plate streaking, Exercise 11

Procedure

Each pair of students is assigned two cultures and a temperature.

Group	Group
A: 63°C _____	A: 72°C _____
B: 63°C _____	B: 72°C _____
C: 63°C _____	C: 72°C _____
D: 63°C _____	D: 72°C _____

You can share beakers of water as long as the effect of the same temperature is being evaluated.

1. Divide two plates of nutrient agar into five sections each. Label the sections "0," "15 sec," "2 min," "5 min," and "15 min."

2. Set up a water bath in the beaker, with the water level higher than the level of the broth in the tubes. Do not put the broth tubes into the water bath at this time. Carefully put the thermometer in a test tube of water in the bath.

3. Streak the assigned organisms on the "0" time section of the appropriate plate. Why are we using "old" and "young" *Bacillus* cultures? _____

_____

4. Raise the temperature of the bath to the desired temperature and maintain that temperature. Use ice to adjust the temperature. Why was 63°C selected as one of the temperatures? _____

_____

5. Place the broth tubes of your organism into the bath when the temperature is at the desired point. After 15 seconds, remove the tubes, resuspend the culture, streak a loopful on the corresponding sections, and return the tubes to the water bath. Repeat at 2, 5, and 15 minutes. What is the longest time period that any microbe is exposed to heat?

_____

6. When you are done, clean the beaker and return the materials. Incubate the plates, inverted, at 35°C until the next lab period. Record your results and the results for the other organisms tested: (−) = no growth; (+) = minimum growth; (2+) = moderate growth; (3+) = heavy growth; and (4+) = maximum growth.

7. Examine the demonstration plates, and record your observations. (Refer to color plate V.6.) Collect results from your classmates to complete the data table in your Laboratory Report.

exercise 22   LABORATORY REPORT

Physical Methods of Control: Heat

Name _____

Date _____

Lab Section _____

Purpose _____

_____

_____

Data

Record growth on a scale from (−) to (4+).

Organism	Temperature/Time									
	63°C					72°C				
	0	15 sec	2 min	5 min	15 min	0	15 sec	2 min	5 min	15 min
Old *Bacillus subtilis*										
Young *Bacillus subtilis*										
Staphylococcus epidermidis										
Escherichia coli										
Mold (*Penicillium*) spores										

Demonstration Plates

	Control	Autoclaved	Dry-Heated
Number of colonies			
Number of different colonies			

Graph your cultures at 63°C.   at 72°C.

Growth 2   Growth 2

Time (min)   Time (min)

Conclusions _____

_____

_____

_____

Questions

1. Compare the heat sensitivity of fungal spores to that of bacterial endospores. _____

   _____

2. Compare the effectiveness of autoclaving and dry heat. _____

   _____

   _____

   _____

3. Give an example of an application (use) of thermal death time. _____

   _____

4. In the exercise, was the thermal death time or thermal death point determined? _____

   _____

5. Give an example of a nonlaboratory use of each of the following methods to control microbial growth:

   a. Incineration: _____

   b. Pasteurization: _____

   c. Autoclaving: _____

6. Define pasteurization. What is the purpose of pasteurization? _____

_____

_____

_____

Critical Thinking

1. Explain why fungi and *Bacillus* sometimes grow better after heat treatment.

2. The decimal reduction time (DRT value) is the time it takes to kill 90% of cells present. Assume that a DRT value for autoclaving a culture is 1.5 minutes. How long would it take to kill all the cells if 10^6 cells were present? What would happen if you stopped the heating process after 9 minutes?

3. Indicators are used in autoclaving to ensure that sterilization is complete. One type of chemical indicator turns color when it has reached a specific temperature; the other type turns color when it has reached a specified temperature and been exposed to steam. Which type of indicator should be used?

4. A biological indicator used in autoclaving is a vial containing 10^9 *Bacillus stearothermophilus* cells that is placed in the autoclave with the material to be sterilized. After autoclaving, the vial is incubated and examined for growth. Why is this species used as opposed to *E. coli* or *B. subtilis*?

exercise 23

Physical Methods of Control: Ultraviolet Radiation

Objectives

After completing this exercise, you should be able to:

1. Examine the effects of ultraviolet radiation on bacteria.
2. Explain the method of action of ultraviolet radiation and light repair of mutations.

Background

Radiant energy comes to the Earth from the Sun and other extraterrestrial sources, and some is generated on Earth from natural and human-made sources. The **radiant energy spectrum** is shown in Figure 23.1. Radiation differs in wavelength and energy. The shorter wavelengths have more energy. X rays and gamma rays are forms of **ionizing radiation.** Their principal effect is to ionize water into *highly reactive free radicals* (with unpaired electrons) that can break strands of DNA. The effect of radiation is influenced by many variables, such as the age of the cells, media composition, and temperature.

Some **nonionizing** wavelengths are essential for biochemical processes. The main absorption wavelengths for green algae, green plants, and photosynthetic bacteria are shown in Figure 23.1a. Animal cells synthesize vitamin D in the presence of light around 300 nm. Nonionizing radiation between 15 and 390 nm is called **ultraviolet (UV).** Wavelengths below 200 nm are absorbed by air and do not reach living organisms. The most lethal wavelengths, sometimes called biocidal, are in the **UVC** range, 200–290 nm. These wavelengths correspond to the optimal absorption wavelengths of DNA (Figure 23.1b). **UVB** wavelengths (290–320 nm) can also cause damage to DNA. **UVA** wavelengths (320–400 nm) are not as readily absorbed and are therefore less active on living organisms.

Ultraviolet light induces *pyrimidine dimers* in the nucleic acid, which result in a mutation. Mutations in critical genes result in the death of the cell unless the damage is repaired. When pyrimidine dimers are exposed to visible light, photolyases are activated; these enzymes split the dimers. This is called **light repair** or **photoreactivation.** Another repair mechanism, called **dark repair,** is independent of light. Dimers are removed by endonuclease, DNA polymerase replaces the bases, and DNA ligase seals the sugar-phosphate backbone.

As a sterilizing agent, ultraviolet radiation is limited by its poor penetrating ability. It is used to sterilize some heat-labile solutions, to decontaminate hospital operating rooms and food-processing areas, and to disinfect wastewater.

In this exercise, we will investigate the effects of ultraviolet radiation and light repair using lamps of the desired wavelength.

Materials

Petri plates containing nutrient agar (3)

Sterile cotton swabs (3)

Covers (choose one): Gauze; 3, 6, or 12 layers of paper; cloth; aluminum foil; clear glass; sunglasses; or plastic

Ultraviolet lamp (265 nm)

Cultures (one of the following)

Serratia marcescens

Bacillus subtilis

Saccharomyces cerevisiae

Techniques Required

Inoculating loop, Exercise 10

Aseptic technique, Exercise 10

Procedure

1. Swab the surface of each plate with *one* of the cultures; to ensure complete coverage, swab the surface in two directions. Label the plates "A," "B," and "C."
2. Remove the lid of an inoculated plate and cover one-half of the plate with one of the covering materials (Figure 23.2). Cover one-half of each of the remaining plates with the same material.

Figure 23.1

Radiant energy. **(a)** Radiant energy spectrum and absorption of light for photosynthesis. **(b)** Biocidal effectiveness of radiant energy between 200 and 700 nanometers (nm) (from UV to visible red light).

Figure 23.2
With the lid removed, cover one-half of an inoculated Petri plate with one of the covering materials.

3. Place each plate directly under the ultraviolet light 24 cm from the light with the *cover off,* agar-side up. Why should the cover be removed? _____

    _____

    Plate A: Expose for 30 seconds. Replace the lid and incubate (22°C).
    Plate B: Expose for 30 seconds and incubate in sunlight. Replace the Petri plate lid and cover the plate with a piece of glass.
    Plate C: Expose for 15 seconds. Replace the lid and incubate in a dark incubator (22°C).

4. Incubate all three plates, inverted, at 22°C or at room temperature until the next period.

5. Examine all plates and record your results. Observe the results of students using the other organisms.

exercise 23

Physical Methods of Control: Ultraviolet Radiation

Name _____

Date _____

Lab Section _____

Purpose _____

_____

_____

Data

What organism did you use? _____

What did you use to cover one-half of each plate? _____

Sketch your results. Note any pigmentation.

A

B

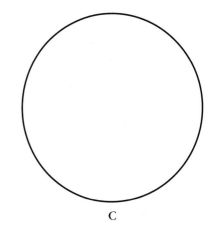

C

Classmates' results from other organisms: _____

_____

_____

Conclusions _____

_____

_____

Questions

1. If the *Bacillus* had sporulated before exposure to radiation, would that affect the results? _____

   _____

   _____

2. What are the variables in ultraviolet radiation treatment? _____

   _____

3. Many of the microorganisms found on environmental surfaces are pigmented. Of what possible advantage is

   the pigment? _____

   _____

   _____

Critical Thinking

1. Can dark repair be a factor in this experiment?

2. Why are there still some colonies growing in the areas exposed to ultraviolet light?

3. How might the results differ if a UVA lamp were used? A UVB lamp?

4. Considering your results, discuss the possible effects of UV radiation on the ecology of a lake if the UV radiation has the same effect on the lake bacteria as it did on the bacteria in your experiment.

exercise 24

Chemical Methods of Control: Disinfectants and Antiseptics

*One nineteenth-century method of avoiding
cholera: Wear a pouch of foul-smelling herbs
around your neck. If the odor is bad enough,
disease carriers will spare you the trouble of avoiding them.*

ANONYMOUS

Objectives

After completing this exercise, you should be able to:

1. Define the following terms: disinfectant and antiseptic.
2. Describe the use-dilution test.
3. Evaluate the relative effectiveness of various chemical substances as antimicrobial agents.

Background

A wide variety of chemicals called **antimicrobial agents** are available for controlling the growth of microbes. Chemotherapeutic agents are used internally and will be evaluated in another exercise. **Disinfectants** are chemical agents used on inanimate objects to lower the level of microbes on their surface; **antiseptics** are chemicals used on living tissue to decrease the number of microbes. Disinfectants and antiseptics affect bacteria in many ways. Those that result in bacterial death are called **bactericidal agents.** Those causing temporary inhibition of growth are **bacteriostatic agents.**

No single chemical is the best to use in all situations. Antimicrobial agents must be matched to specific organisms and environmental conditions. Additional variables to consider in selecting an antimicrobial agent include pH, solubility, toxicity, organic material present, and cost. In evaluating the effectiveness of antimicrobial agents, the concentration, length of contact, and whether it is lethal (*-cidal*) or inhibiting (*-static*) are the important criteria. The standard method for measuring the effectiveness of a chemical agent is the **American Official Analytical Chemist's use-dilution test.** For most purposes, three strains of bacteria are used in this test: *Salmonella choleraesuis, Staphylococcus aureus*, and *Pseudomonas aeruginosa*. To perform a use-dilution test, metal rings are dipped into standardized cultures of the test bacteria grown in liquid media, removed, and dried. The rings are then placed into a solution of the disinfectant at the concentration recommended by the manufacturer for 10 minutes at 20°C. The rings are then transferred to a nutrient medium to permit the growth of any surviving bacteria. The effectiveness of the disinfectant can then be determined by the amount of resulting growth. The use-dilution test is limited to bactericidal compounds and cannot be used to evaluate bacteriostatic compounds.

In this exercise, we will perform a modified use-dilution test.

Materials

Petri plates containing nutrient agar (2)

Sterile water

Sterile tubes (3)

5 ml sterile pipettes (2)

1 ml sterile pipettes (2)

Test substance: chemical agents such as bathroom cleaner, floor cleaner, mouthwash, lens cleaner, acne cream. Bring your own.

Culture

Staphylococcus aureus

Techniques Required

Inoculating loop, Exercise 10

Aseptic technique, Exercise 10

Pipetting, Appendix A

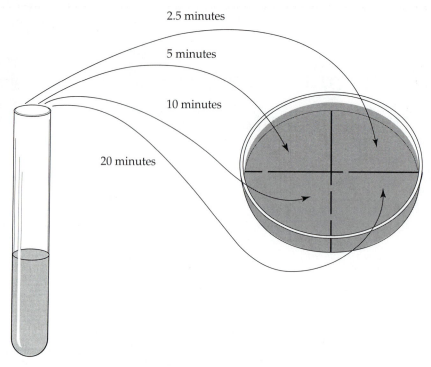

Chemical and bacteria

Figure 24.1
Transfer a loopful from the tube containing the lab disinfectant and *Staphylococcus* to the appropriate sector at the time intervals shown. Repeat the procedure with a loopful from the tube containing the test substance and *Staphylococcus* onto the appropriate sector of the second nutrient agar plate.

Procedure

1. Using sterile water, prepare a dilution of the test substance in a sterile tube, diluted to the strength at which it is normally used. If it is a paste, it must be suspended in sterile water.

2. Transfer 5 ml of the diluted test substance to a sterile tube. Label the tube. Add 5 ml of your laboratory disinfectant to another sterile tube. What is the disinfectant you use to disinfect your lab bench? _____ _____ Label the tube.

3. Divide one plate of nutrient agar into five sections. Label the sections "0," "D-2.5," "D-5," "D-10," and "D-20." The D stands for laboratory disinfectant.

4. Label the other nutrient agar plate for the other chemical and divide it into four sections. Label the sections "2.5," "5," "10," and "20."

5. Inoculate the 0 sector with a loopful of *S. aureus*.

6. Aseptically add 0.5 ml of the *S. aureus* culture to each tube prepared in step 2.

7. Transfer one loopful from each tube to a corresponding sector at 2.5 minutes, 5 minutes, 10 minutes, and 20 minutes (Figure 24.1).

8. Incubate the plates, inverted, at 35°C until the next lab period. (Discard the chemical/bacteria mixtures in the To Be Autoclaved area.)

9. Observe the plates for growth. Record the growth as $(-)$ = no growth; $(+)$ = minimum growth; $(2+)$ = moderate growth; $(3+)$ = heavy growth; and $(4+)$ = maximum growth.

exercise 24

Chemical Methods of Control: Disinfectants and Antiseptics

Name _____

Date _____

Lab Section _____

Purpose _____

_____

_____

Data

Time of Exposure (min)	Amount of Growth		
	Control	Lab Disinfectant	Chemical: _____
0			
2.5			
5			
10			
20			

Conclusions _____

_____

_____

Questions

1. Was this a fair test? Is it representative of the effectiveness of the test substance? _____

_____

_____

2.   Read the label of the preparation you tested. What is (are) the active ingredient(s)? _____

_____

Using your textbook or another reference, find the method of action of the test substance. _____

_____

_____

_____

3.   What is the use-dilution method? _____

_____

_____

Critical Thinking

1.   How could the procedures used in this experiment be altered to measure bacteriostatic effects?

2.   In the use-dilution test, a chemical is evaluated by its ability to kill 10^6 to 10^8 dried *Clostridium sporogenes* or *Bacillus subtilis* endospores. Why is this considered a stringent test?

3.   The effectiveness of disinfectants can be measured in DRT values. DRT, or decimal reduction time, is the length of time it takes to kill 90% of a test population of bacteria. The DRT values for contact lens disinfectants against *Serratia marcescens* are as follows:

Disinfectant	DRT Value (min)	Disinfectant	DRT Value (min)
Chlorhexidine, 0.005%	2.8	Thimerosal, 0.002%	138.9
Hydrogen peroxide, 3%	3.1	Polyquaternium-1, 0.001%	383.3

Which disinfectant is most effective? _____ What is the minimum time that lenses with

10^2 bacteria should be soaked in chlorhexidine? _____

In polyquaternium-1? _____ What if the lenses are contaminated with *Staphylococcus* or

Acanthamoeba? _____

_____

Why isn't a higher concentration of disinfectant used? _____

exercise 25

Chemical Methods of Control: Antimicrobial Drugs

*The aim of medicine is to prevent disease and prolong life;
the ideal of medicine is to eliminate the need of a physician.*

WILLIAM JAMES MAYO

Objectives

After completing this exercise, you should be able to:

1. Define the following terms: antibiotic, chemotherapeutic agent, and MIC.
2. Perform an antibiotic sensitivity test.
3. Provide the rationale for the agar diffusion technique.

Background

The observation that some microbes inhibited the growth of others was made as early as 1874. Pasteur and others observed that infecting an animal with *Pseudomonas aeruginosa* protected the animal against *Bacillus anthracis*. Later investigators coined the word **antibiosis** (against life) for this inhibition and called the inhibiting substance an **antibiotic.** In 1928, Alexander Fleming observed antibiosis around a mold (*Penicillium*) growth on a culture of staphylococci. He found that culture filtrates of *Penicillium* inhibited the growth of many gram-positive cocci and *Neisseria* spp. In 1940, Selman A. Waksman isolated the antibiotic streptomycin, produced by an actinomycete. This antibiotic was effective against many bacteria that were not affected by penicillin. Actinomycetes remain an important source of antibiotics. Today, research investigators look for antibiotic-producing actinomycetes and fungi in soil, and have synthesized many antimicrobial substances in the laboratory. Antimicrobial chemicals absorbed or used internally, whether natural (antibiotics) or synthetic, are called **chemotherapeutic agents.**

A physician or dentist needs to select the correct chemotherapeutic agent intelligently and administer the appropriate dose in order to treat an infectious disease; then the practitioner must follow that treatment in order to be aware of resistant forms of the organism that might occur. The clinical laboratory isolates the **pathogen** (disease-causing organism) from a clinical sample and determines its sensitivity to chemotherapeutic agents.

In the **disk-diffusion method,** a Petri plate containing an agar growth medium is inoculated uniformly over its entire surface. Paper disks impregnated with various chemotherapeutic agents are placed on the surface of the agar. During incubation, the chemotherapeutic agent *diffuses* from the disk, from an area of high concentration to an area of lower concentration. An effective agent will inhibit bacterial growth, and measurements can be made of the size of the **zones of inhibition** around the disks. The concentration of chemotherapeutic agent at the edge of the zone of inhibition represents its **minimum inhibitory concentration (MIC).** The MIC is determined by comparing the zone of inhibition with MIC values in a standard table (Table 25.1). The zone size is affected by such factors as the diffusion rate of the antibiotic and the growth rate of the organism. To minimize the variance between laboratories, the standardized **Kirby–Bauer test** for agar diffusion methods is performed in many clinical laboratories. This test uses *Mueller-Hinton agar*. Mueller-Hinton agar allows the chemotherapeutic agent to diffuse freely.

In this exercise, we will evaluate chemotherapeutic agents by the disk-diffusion method.

Materials

Petri plate containing Mueller-Hinton agar

Sterile cotton swab

Dispenser and antimicrobial disks

Forceps

Alcohol

Ruler (second period)

Cultures (as assigned)

Staphylococcus aureus

Escherichia coli

Pseudomonas aeruginosa

Table 25.1

Interpretation of Inhibition Zones of Test Cultures*

Disk Symbol	Chemotherapeutic Agent	Disk Content	Resistant	Intermediate	Susceptible
				Diameter of Zones of Inhibition (mm)	
AM	Ampicillin when testing gram-negative bacteria	10 µg	<13	14–16	>17
AM	Ampicillin when testing gram-positive bacteria	10 µg	<28	—	>29
ATM	Aztreonam	30 µg	<15	16–21	>22
B	Bacitracin	10 units	<8	9–12	>13
C	Chloramphenicol	30 µg	<12	13–17	>18
CAZ	Ceftazidime	30 µg	<14	15–17	>18
CIP	Ciprofloxacin	5 µg	<15	16–20	>21
E	Erythromycin	15 µg	<13	14–22	>23
G	Sulfisoxazole (Gantrisin)	300 µg	<12	13–16	>17
IPM	Imipenem	10 µg	<13	14–15	>16
P	Penicillin G when testing staphylococci	10 units	<28	—	>29
P	Penicillin G when testing other bacteria	10 units	<14	—	>15
S	Streptomycin	10 µg	<11	12–14	>15
SxT	Trimethoprim/Sulfamethoxazole	1.25 µg/ 23.75 µg	<10	11–15	>16
TE	Tetracycline	30 µg	<14	15–18	>19
VA	Vancomycin	30 µg	<9	10–11	>12
VA	Vancomycin when testing enterococci	30 µg	<14	15–16	>17

*National Committee for Clinical Laboratory Standards. Performance Standards for Antimicrobial Disk Susceptibility Tests.

Techniques Required

Inoculating loop, Exercise 10

Aseptic technique, Exercise 10

Procedure

1. Aseptically swab the assigned culture onto the appropriate plate. Swab in three directions to ensure complete plate coverage (Figure 25.1). Why is complete coverage essential? _____
    _____

Let stand at least 5 minutes.

2. Follow procedure a or b.
    a. Place the chemotherapeutic-impregnated disks by pushing the dispenser over the agar. Sterilize your loop and touch each disk with a sterile inoculating loop to ensure better contact with the agar. Record the agents and the disk codes in your Laboratory Report. Circle the corresponding chemicals in Table 25.1.
    b. Sterilize forceps by dipping them in alcohol and burning off the alcohol.

Area of initial swab   Area of second swab

Figure 25.1
Dip a cotton swab in the culture to be tested and swab across the surface of the agar without leaving any gaps. Using the same swab, swab the agar in a direction perpendicular to the first inoculum. Repeat, swabbing the agar at a 45° angle to the first inoculum.

> ⚠ **Keep the beaker of alcohol away from the flame.**

Obtain a disk impregnated with a chemotherapeutic agent and place it on the surface of the agar (Figure 25.2a). Gently tap the disk with a sterile inoculating loop to ensure better contact with the agar. Repeat, placing five to six different disks the same distance apart on the Petri plate. See the location of the disks in Figure 25.2b. Record the agents and the disk codes in your Laboratory Report. Circle the corresponding chemicals in Table 25.1.

3. Incubate the plate, inverted, at 35°C until the next period. Measure the zones of inhibition in millimeters, using a ruler on the underside of the plate (Figure 25.2b). Record the zone size and, based on the values in Table 25.1, indicate whether the organism is sensitive, intermediate, or resistant. Observe the results of students using the other two bacteria. (See color plates VIII.1 and VIII.2.)

(a) Place disks impregnated with chemotherapeutic agents on an inoculated culture medium with sterile forceps to get the pattern shown in **(b)**.

(b) After incubation, measure the diameters of zones of inhibition.

Figure 25.2
Disk-diffusion method.

exercise 25

Chemical Methods of Control: Antimicrobial Drugs

Name _____

Date _____

Lab Section _____

Purpose _____

_____

_____

Data

Antimicrobial Agent	Disk Code	Staphylococcus aureus		Escherichia coli		Pseudomonas aeruginosa	
		Zone Size	S, I, or R*	Zone Size	S, I, or R*	Zone Size	S, I, or R*
1.							
2.							
3.							
4.							
5.							
6.							
7.							
8.							

*S = susceptible; I = intermediate; R = resistant.

Conclusions

Which chemotherapeutic agent was most effective against each organism? _____

_____

_____

Questions

1. Is the disk-diffusion technique measuring bacteriostatic or bactericidal activity? Briefly explain.

_____

_____

2. In which growth phase is an organism most sensitive to an antibiotic? _____

_____

3. Why is the disk-diffusion technique not a perfect indication of how the drug will perform in vivo? What other factors are considered before using the chemotherapeutic agent in vivo? _____

_____

_____

4. Using your textbook or other references, match each of the antimicrobial agents listed in Table 25.1 with its type and method of action.

Type of Antimicrobial Agent
a. Aminoglycosides
b. β-lactams
c. Carbapenems
d. Cephalosporins
e. Glycopeptides
f. Macrolides
g. Monobactams
h. Quinolones
i. Sulfonamides
j. Tetracyclines
k. None of the above

Method of Action
1. Enzyme inhibitor
2. Inhibit cell wall synthesis
3. Inhibit protein synthesis

	Type	Method of Action
Ampicillin		
Aztreonam		
Bacitracin		
Ceftazidime		
Chloramphenicol		
Ciprofloxacin		
Erythromycin		
Imipenem		
Penicillin G		
Streptomycin		
Sulfisoxazole		
Tetracycline		
Trimethoprim/Sulfamethoxazole		
Vancomycin		

5. From your results, which type of antimicrobial agent was most effective against gram-negative bacteria? _____

_____

Against gram-positive bacteria? _____

Critical Thinking

1. What effect would the presence of tetracycline in the body have on penicillin therapy?

2. The following results were obtained from a disk-diffusion test against a bacterium:

Antibiotic	Zone of Inhibition (mm)
A	0
B	12
C	5
D	12

Which drug should be used to treat an infection caused by this bacterium? Briefly explain.

3. The broth dilution test can be used to determine the effectiveness of an antibiotic. In this test, serial dilutions of the antibiotic were set up in the wells of a microtiter plate. Equal amounts of broth culture of *Staphylococcus aureus* were added to each well. After incubation, the wells were examined for bacterial growth. Wells with no growth were subcultured in nutrient broth without the antibiotic. Results were recorded as (+) for growth and (−) for no growth.

Antibiotic	Dilution	Growth	Growth in Subculture
A	1:10 through 1:70	−	−
	1:80	−	−
	1:90	−	+
	1:100	−	+
	1:200 through 1:500	+	+
B	1:10 through 1:150	−	−
	1:160	−	+
	1:170	+	+
	1:180	+	+
	1:190 through 1:500	+	+

What is the minimum bactericidal concentration of each antibiotic? _____

What is the minimum bacteriostatic concentration? _____

Which antibiotic is more effective against S. *aureus*? _____

Which antibiotic is more effective against S. *typhimurium*? _____

exercise 26

Effectiveness of Hand Scrubbing

People are sick because they are poor, they become poorer because they are sick and they become sicker because they are poorer.

ANONYMOUS

Objectives

After completing this exercise, you should be able to:

1. Evaluate the effectiveness of hand washing and a surgical scrub.
2. Explain the importance of aseptic technique in the hospital environment.

Background

The skin is sterile during fetal development. After birth, a baby's skin is colonized by many bacteria for the rest of its life. As an individual ages and changes environments, the microbial population changes to match the environmental conditions. The microorganisms that are more or less permanent are called **normal microbiota.** Microbes that are present only for days or weeks are referred to as **transient microbiota.**

Discovery of the importance of hand washing in disease prevention is credited to Ignaz Semmelweis at Vienna General Hospital in 1846. He noted that the lack of aseptic methods was directly related to the incidence of puerperal fever and other diseases. Medical students would go directly from the autopsy room to the patient's bedside and assist in child delivery without washing their hands. Less puerperal sepsis occurred in patients attended by midwives, who did not touch cadavers. Semmelweis established a policy for the medical students of hand washing with a chloride of lime solution that resulted in a drop in the death rate due to puerperal sepsis from 12% to 1.2% in one year. Guidelines from the Centers for Disease Control and Prevention state that "handwashing is the single most important procedure for preventing nosocomial infections," yet recent studies in hospitals show handwashing rates as low as 31%.

A layer of oil and the structure of the skin prevent the removal of all bacteria by hand washing. Soap helps remove the oil, and scrubbing will maximize the removal of bacteria. Hospital procedures require personnel to wash their hands before attending a patient, and a complete surgical scrub—removing the transient and

many of the resident microbiota—is done before surgery. Transient microbiota are usually removed after 10 to 15 minutes of scrubbing with soap. The surgeon's skin is never sterilized. Only burning or scraping it off would achieve that.

In this exercise, we will examine the effectiveness of washing skin with soap and water. Only organisms capable of growing aerobically on nutrient agar will be observed. Because organisms with different nutritional and environmental requirements will not grow, this procedure will involve only a minimum number of the skin microbiota.

Materials

Petri plates containing nutrient agar (2)

Scrub brush

Bar soap or liquid soap (bring one from home)

Techniques Required

Colony morphology, Exercise 9

Procedure

1. Divide two nutrient agar plates into four quadrants. Label the sections of each plate 1 through 4. Label one plate "Water" and the other "Soap." Which soap will you use? _____
2. Do the "Water" plate first. Touch section 1 with your fingers, wash well *without* soap, shake off excess water, and, while your hands are still wet, touch section 2. Do not dry your fingers with a towel. Wash again, and, while your hands are still wet, touch section 3. Wash a final time, and touch section 4.
3. Use your other hand on the plate labeled "Soap." Wash with soap, rinse, shake off the excess water, then touch section 1.
4. Wash again with soap, rinse, shake off the excess water, then touch section 2.

5. Using a brush and soap, scrub your hand for 2 minutes, rinse, shake off the excess water, then touch section 3.
6. Repeat the soap-and-brush scrub for 4 minutes, rinse, shake off the excess water, then touch section 4.
7. Incubate the plates, inverted, at 35°C until the next period.

8. Speculate on your expected results, and record them in your Laboratory Report.
9. Record the growth as (−) = no growth; (+) = minimum growth; (2+) = moderate growth; (3+) = heavy growth; and (4+) = maximum growth.

exercise 26

Effectiveness of Hand Scrubbing

Name _____

Date _____

Lab Section _____

Purpose

_____

_____

_____

Expected Results

Before the next lab period, indicate the relative amounts of growth you *expect* in each quadrant.

Section	Water Alone	Soap
1.	(No washing)	
2.		
3.		
4.		

Data

Indicate the relative amounts of growth in each quadrant.

Section	Water Alone	Soap (type: liquid pink soap)
1.	(No washing) several sm & large colonies	lots of sm colonies white round
2.	two large circular a couple sm	very few sm. round
3.	one very big	blob looking yellow and opacte
4.	one medium several sm round colonies	several very sm. colonies

Conclusions

without or with soap you still have bacteria on our hands even w/ Hand Sanitizaton.

Questions

1. Did your results differ from your expected results? _____

    Briefly explain why or why not. _____

2. What is a surgeon trying to accomplish with a 10-minute scrub with a brush followed by an antiseptic? _____

    _____

    _____

3. How do normal microbiota and transient microbiota differ? _____

    _____

4. Using your classmates' data, compare the results from bar soap and liquid soap. _____

    _____

    _____

Critical Thinking

1. If most of the normal microbiota and transient microbiota aren't harmful, then why must hands be scrubbed before surgery?

2. The following data were collected from soaps after one week of use at a hospital nurses' hand-washing station. Neither bacteria nor fungi were isolated from any of the products before use.

Isolation of aerobic bacteria from 25 soap products. Data are expressed as percentage of soap products contaminated.

Organisms	Bar Soap	Liquid Soap: Type of Closure			
		Screw top	Slit/flip	Flip/pump	Pump
Total bacteria	95%	71%	39%	10%	0%
Gram-positive cocci	95%	71%	39%	10%	0%
Gram-negative rods	12%	1%	1%	1%	0%

What conclusions can you draw from these data?

part seven

Microbial Genetics

EXERCISES

All of the characteristics of bacteria—including growth patterns, metabolic activities, pathogenicity, and chemical composition—are inherited. These traits are transmitted from parent cell to offspring through genes. **Genetics** is the study of genes: how they carry information, how they are replicated and passed to the next generation or to another organism, and how their information is expressed within an organism to determine the particular characteristics of that organism (see the illustration on the next page).

The information stored in genes is called the **genotype.** All of the genes may not be expressed. **Phenotype** refers to the actual expressed characteristics, such as the ability to perform certain biochemical reactions (e.g., synthesis of a capsule or pigment).

Bacteria are invaluable to the study of genetics because large numbers can be cultured inexpensively, and their relatively simple genetic composition facilitates studying the structure and function of genes. The control of gene expression is explained by the operon model (Exercise 27). This model was first proposed by François Jacob and Jacques Monod to explain why *E. coli* makes lactose-catabolizing enzymes in the presence, but not in the absence, of lactose.

Bacteria undergo genetic change due to mutation or recombination. These genetic changes result in a change in the genotype. When a change occurs, in most instances we can expect that the product encoded by certain genes will be changed. An enzyme encoded by a changed gene may become inactive. This might be disadvantageous or even lethal if the cell loses a phenotypic trait it needs. Some genetic changes may be beneficial.

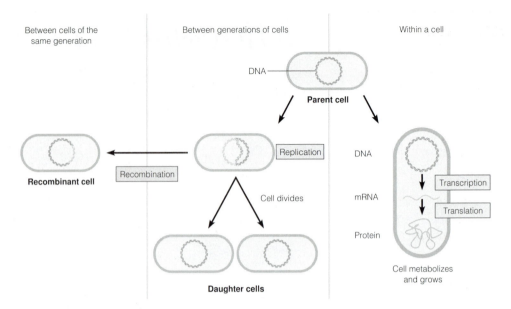

Between cells of the
same generation

Between generations of cells

Within a cell

DNA

Parent cell

Replication

Recombination

Recombinant cell

Cell divides

Daughter cells

DNA

mRNA

Protein

Transcription

Translation

Cell metabolizes
and grows

The flow of genetic information. Genetic information can be transferred between generations of cells, through DNA replication. Occasionally, genetic information can be transferred between cells of the same generation, through recombination. Genetic information is also used within a cell to produce the proteins the cell needs to function, through transcription and translation.

For instance, if an altered enzyme encoded by a changed gene has new enzymatic activity, the cell may be able to grow in new environments.

Genetic change in microbial populations is relatively easy to observe. In 1949, Howard B. Newcombe wrote:

> *Numerous bacterial variants are known which will grow in environments unfavorable to the parent strain, and to explain their occurrence two conflicting hypotheses have been advanced. The first assumes that the particular environment produces the observed change in some bacteria exposed to it, whereas the second assumes that the variants arose spontaneously during growth under normal conditions.*[*]

Salvador Luria and Max Delbrück first demonstrated the latter **spontaneous mutation** hypothesis in 1943.

Bacteria can be mutated to get a different product from an altered gene (Exercise 28). New genes can be inserted into bacteria by transformation (Exercise 29). The desired gene can be isolated and identified by DNA fingerprinting (Exercise 30). Artificial manipulation of genes is known as **genetic engineering.** In genetic engineering, genes can be inserted into a vector and incorporated into a bacterium by transformation (Exercise 31).

Cancer is the result of *permanent changes* in the nucleotide sequence of chromosomes. We will perform a test using principles of microbial genetics for identification of possible cancer-inducing substances in Exercise 32.

[*]Quoted in H. A. Lechevalier and M. Solotorovsky. *Three Centuries of Microbiology.* New York: Dover Publications, 1974, p. 502.

exercise 27

Regulation of Gene Expression

Objectives

After completing this exercise, you should be able to:

1. Define operon, induction, and repression.
2. Determine the inducers for the enzymes tested in this exercise.

Background

Many genes are expressed constantly throughout the life of a cell. However, some proteins may be needed only during a particular growth phase or in a particular environment. A cell can conserve energy by making only those proteins needed at a particular time.

The operon model was proposed to explain why *Escherichia coli* made the enzyme β-galactosidase in the presence of lactose but not in its absence. An operon consists of a promoter, the region where RNA polymerase binds; the operator, which signals whether transcription will occur; and the structural genes they control. The structural genes code for the peptides. A regulatory gene encodes a regulatory protein, which binds to the operator to activate or inhibit transcription. In an **inducible operon,** transcription is activated or induced by the presence of a particular substance—in this case, lactose. The regulatory protein has an allosteric site to which the **inducer** can bind. The inducer then inactivates the repressor. A **repressible operon** is inhibited when a specific small

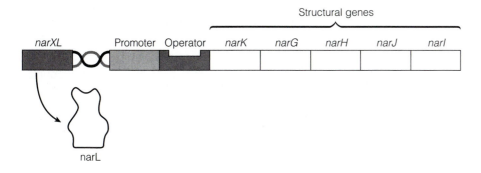

(a) The *nar* operon. The regulatory gene (*narXL*) encodes the regulatory protein narL.

(b) The *ara* operon. The regulatory gene (*araC*) encodes the regulatory protein araC.

Figure 27.1
An operon consists of a promoter, an operator, and the structural genes they control.

Regulatory
gene

Promoter Operator

Structural genes

Regulatory
protein

(a) A regulatory gene codes for a regulatory protein.

RNA
polymerase

(b) The regulatory protein binds to the operator and prevents transcription of the structural
genes.

(c) A change in shape of the repressor protein activates transcription of the structural genes.

Figure 27.2
The regulatory proteins that control the *nar* and *ara* operons can inhibit or activate transcription.

molecule is present. The synthesis of the amino acid tryptophan in *E. coli* is inhibited by tryptophan. If tryptophan binds to the regulatory protein, the tryptophan-protein complex acts as a repressor to inhibit synthesis of more tryptophan.

In *E. coli* the genes for nitrate reductase are encoded by the **nar operon** (Figure 27.1a), and the enzymes for the catabolism of arabinose are encoded by the **ara operon** (Figure 27.1b). The regulatory genes for these operons produce a protein that binds to the operator to inhibit transcription or to promote binding of RNA polymerase, depending on the shape of the regulatory proteins (Figure 27.2).

In this experiment, the *ara* structural genes have been replaced with a gene called pGLO™* that encodes Green Fluorescent Protein (see color plate II.4). The pGLO gene, isolated from the jellyfish *Aequoria victoria*, is widely used to study gene expression. You will investigate the *ara* or *nar* operons in *E. coli* in this exercise. Before beginning, write your hypothesis in your Laboratory Report.

*pGLO is a trademark of Bio-Rad Laboratories.

Materials

Small test tubes (2)

1 ml pipettes (3)

Nitrate solution

Nitrate reagents A and B

Brewer anaerobic jar

Petri plate containing nutrient agar

Petri plate containing arabinose–nutrient agar

Spreading rod

Alcohol

Ultraviolet lamp

Cultures

Escherichia coli for nitrate reductase

Escherichia coli pGLO for arabinose

Techniques Required

Anaerobic culture techniques, Exercise 19

Nitrate reduction test, Exercise 17

Pipetting, Appendix A

Procedure

Nitrate Reductase

1. Label two small tubes "Aerobic" and "Anaerobic." Add 0.5 ml *E. coli* to each tube. Then add 0.2 ml nitrate (NO_3^-) solution to each tube.
2. Immediately place the "anaerobic" tube in the anaerobic Brewer jar and activate the Gas-Pak®. Begin shaking the "aerobic" tube. Do not splash the contents out of the tube, but shake it enough to aerate the entire culture.
3. After 15 minutes, test for the presence of nitrite ions (NO_2^-) by adding 5 drops of nitrate reagent A and 5 drops of nitrate reagent B to each tube. Be sure to test an uninoculated nitrate broth tube for nitrate ion. Let the tubes stand at room temperature for 15 to 20 minutes. The presence of a pink color indicates nitrite ions. Observe the tubes for a color change and record your results.

Figure 27.3
Inoculation using a spreading rod.

Arabinose Catabolism

1. Aseptically transfer 0.1 ml of the *E. coli* pGLO culture to the surface of the nutrient agar and the arabinose–nutrient agar.
2. Disinfect a spreading rod by dipping it in alcohol, quickly igniting the alcohol in a Bunsen burner flame, and letting the alcohol burn off.

 While it is burning, hold the spreading rod pointed *down*. Keep the beaker of alcohol away from the flame.

Let the spreading rod cool.
3. Spread the liquid on the surface of each plate. Let the Petri plates sit undisturbed until the liquid diffuses into the agar (Figure 27.3).

4. Incubate the plates in an inverted position at 35°C for 24 to 48 hours.
5. Observe the plates using an ultraviolet lamp. Are any of the colonies producing a fluorescent protein? _____

 Do not look at the ultraviolet light, and do not leave your hand exposed to it.

---

exercise 27

Regulation of Gene Expression

Name _____

Date _____

Lab Section _____

Purpose _____

_____

_____

Hypothesis _____

_____

_____

Results

Nitrate Reductase

	Color After NO_3^- Reagents	NO_3^- Reduction?
Control		
E. coli, aerobic		
E. coli, anaerobic		

Arabinose Catabolism

	Growth	Fluorescence
Nutrient agar		
Arabinose–nutrient agar		

Conclusions

Did you accept or reject your hypothesis? Briefly explain. _____

_____

_____

_____

Questions

1. Write the chemical reaction catalyzed by nitrate reductase.

2. What is the inducer for *nar*? _____

3. Of what value is nitrate reductase to *E. coli*? _____

_____

_____

4. Chemically what is arabinose? _____

5. What is the inducer for *ara*? _____

6. *E. coli* do not normally fluoresce in the presence of arabinose. What do they normally do? _____

_____

_____

7. Why did the *E. coli* cells fluoresce in the presence of arabinose in this experiment? _____

_____

_____

    Why not in the absence of arabinose? _____

8. Why might some *E. coli* cells not fluoresce? _____

_____

_____

Critical Thinking

1. Several bacteria including *Bacillus subtilis* reduce nitrate to ammonia as the first step in amination. This is called assimilatory nitrate reduction. *E. coli*'s transformation of nitrate is called dissimilatory nitrate reduction. How do assimilatory and dissimilatory nitrate reduction differ?

2. What is the valence of N in NO_3^-? In NO_2^-? What is gaining electrons in nitrate reduction?

3. How do you expect the presence of glucose to affect the arabinose operon? Briefly explain.

4. Assume you replaced the *ara* gene with the gene for amylase in the *ara* operon. What would be produced in

    the presence of arabinose? _____

exercise 28

Isolation of Bacterial Mutants

Though coli may bother and vex us,
It's hard to believe they outsex us.
They accomplish seduction
by viral transduction
Foregoing the joys of amplexus.

SEYMOUR GILBERT

Objectives

After completing this exercise, you should be able to:

1. Define the following terms: prototroph, auxotroph, and mutation.
2. Differentiate between direct and indirect selection.
3. Isolate bacterial mutants by replica plating.

Background

For practical purposes, genes and the characteristics for which they code are stable. However, when millions of bacterial progeny are produced in just a few hours of incubation, genetic variants may occur. A variant is the result of a **mutation**—that is, a change in the sequence of nucleotide bases in the cell's DNA.

Metabolic mutants can be easily identified and isolated. Catabolic mutations could result in a bacterium that is deficient in the production of an enzyme needed to utilize a particular substrate. Anabolic mutations result in bacteria that are unable to synthesize an essential organic chemical. "Wild-type," or nonmutated, bacteria are called **prototrophs,** and mutants are called **auxotrophs.** Auxotrophs have been isolated that either cannot catabolize certain organic substrates or cannot synthesize certain organic chemicals, such as amino acids, purine or pyrimidine bases, or sugars.

The cultures used in this exercise are capable of synthesizing all of their growth requirements from glucose–minimal salts medium (Table 28.1). Mutations will be induced by exposing the cells to the mutagenic wavelengths of ultraviolet light (Exercise 23.) After exposure to ultraviolet radiation, auxotrophs that cannot grow on glucose–minimal salts medium will be identified.

Because of the low rate of appearance of recognizable mutants, special techniques have been developed to select for desired mutants. In the **gradient plate** (Exercise 21), **direct selection** is used to pick out mutant cells while rejecting the unmutated parent cells. We

Table 28.1
Composition of Glucose–Minimal Salts Agar

Glucose	0.1 g
Dipotassium phosphate	0.7 g
Monopotassium phosphate	0.2 g
Sodium citrate	0.05 g
Magnesium sulfate	0.01 g
Ammonium sulfate	0.15 g
Agar	1.5 g
Water	100 ml

will use **indirect selection** with the **replica-plating technique** in this exercise.

In replica plating, mutated bacteria are grown on a nutritionally complete solid medium. An imprint of the colonies is made on velveteen-covered or rubber-coated blocks and transferred to glucose–minimal salts solid medium and to a complete solid medium. Colonies that grow on the complete medium but not on the minimal medium are auxotrophs. Auxotrophs are identified *indirectly* because they will not grow.

Materials

Petri plates containing nutrient agar (complete medium) (3)

99 ml water dilution blanks (3)

Sterile 1 ml pipettes (4)

Ultraviolet lamp, 260 nm

Spreading rod

Alcohol

Second period:

Petri plate containing glucose–minimal salts agar (minimal medium)

Petri plate containing nutrient agar (complete medium)

Sterile replica-plating block

Culture

Serratia marcescens

Techniques Required

Spreading rod, Exercise 27

Pipetting, Appendix A

Serial dilution technique, Appendix B

Procedure*

First Period

1. Label the nutrient agar plates "A," "B," and "C" and the dilution blanks "1," "2," and "3."

*Adapted from C. W. Brady. "Replica Plate Isolation of an Auxotroph." Unpublished paper. Whitewater, WI: University of Wisconsin, n.d.

2. Aseptically pipette 1 ml of the assigned broth culture to dilution blank 1 and mix well.
3. Using another pipette, transfer 1 ml from dilution blank 1 to dilution blank 2, as shown in Figure 28.1, and mix well. What dilution is in bottle 2?

    _____

    _____

4. Using another pipette, transfer 1 ml from dilution blank 2 to bottle 3 and 1 ml to the surface of plate A.
5. Mix bottle 3, and transfer 1 ml to the surface of plate B with a sterile pipette and 0.1 ml to the surface of plate C.
6. Disinfect a spreading rod by dipping it in alcohol, quickly igniting the alcohol in a Bunsen burner flame, and letting the alcohol burn off.

> ⚠ **While it is burning, hold the spreading rod pointed *down*. Keep the beaker of alcohol away from the flame.**

Let the spreading rod cool.

7. Spread the liquid on the surface of plate C over the entire surface. Do the same with plates B and A (Figure 27.3). Let the Petri plates sit undisturbed until the liquid diffuses into the agar.

Figure 28.1
Dilution procedure.

8. Disinfect the spreading rod and return it. Why is it not necessary to disinfect the hockey stick between each plate when proceeding from plate C to B to A? _____

_____

9. Put the plate with the cover off, agar-side up, about 30 cm from the ultraviolet lamp. Position the plate directly under the lamp. Turn the lamp on for 30 to 60 seconds, as assigned by your instructor.

> ⚠ **Do not look directly at the light, and do not expose your hand to it.**

10. Incubate the plates at 25°C until the next period.

Second Period

1. Select a plate with 25 to 50 isolated colonies. Mark the bottom of the plate with a reference mark (Figure 28.2a). This is the master plate.
2. Mark the uninoculated complete and minimal media with a reference mark on the bottom of each plate.
3. If your replica-plating block is already assembled, proceed to step 4. If not, follow these instructions. Carefully open the package of velveteen. Place the replicator block on the center of the velveteen. Pick up the four corners of the cloth and secure them tightly on the handle with a rubber band.

4. Inoculate the sterile media by either step a or step b, as follows:
  a. Hold the replica-plating block by resting it on the table with the rough surface, or velveteen, up (Figure 28.2b). Invert the master plate selected in step 1 on the block, and allow the master plate agar to lightly touch the block. Remove the cover from the minimal medium, align the reference marks, and touch the uninoculated minimal agar with the inoculated replica-plating block. Replace the cover. Remove the cover from the uninoculated complete medium and inoculate with the replica-plating block, keeping the reference marks the same.
  b. Place the master plate and the uninoculated plates on the table. Place the reference marks in a 12 o'clock position and remove the covers. Touch the replica-plating block to the master plate; then, without altering its orientation, gently touch it to the minimal medium, then to the complete medium (Figure 28.2c).
5. Replace the covers, and incubate as before. Refrigerate the master plate.
6. After incubation, compare the plates and record your results. (See color plate V.I.)

Figure 28.2
Replica plate. **(a)** Schematic diagram. **(b)** Transfer with stationary replica-plating block. **(c)** Transfer by moving the latex transfer surface from plate to plate.

exercise 28

Isolation of Bacterial Mutants

Name _____

Date _____

Lab Section _____

Purpose _____

_____

_____

Data

Mark the location of colonies, and note any changes in pigmentation.

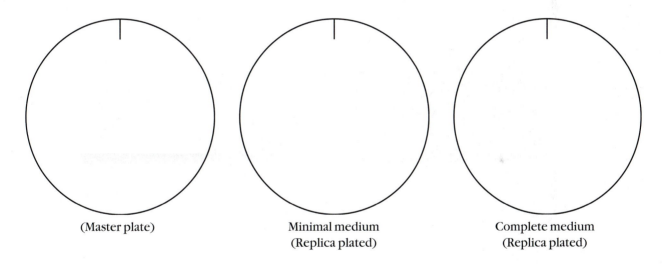

(Master plate)   Minimal medium   Complete medium
                        (Replica plated)   (Replica plated)

Class Data

S. marcescens	UV Exposure (sec)	Number of Auxotrophs

Conclusions

What is the effect of longer exposure to ultraviolet light? _____

_____

_____

Questions

1. Circle the auxotrophs on the diagram of the complete medium.

2. Why was the complete medium, as well as the minimal medium, replica plated? _____

_____

_____

3. Why is replica plating usually used for indirect selection of mutants? _____

_____

_____

Critical Thinking

1. How would you isolate a mutant from the replica plate? Does your technique involve a catabolic or anabolic mutant?

2. Why do you think bacteria such as *Staphylococcus aureus* that survive on surfaces are often pigmented?

3. How would you identify the growth factor(s) needed by an auxotroph?

4. How would you use the replica-plating technique to isolate a mutant that is resistant to an antibiotic?

exercise 29

Transformation of Bacteria

Objectives

After completing this exercise, you should be able to:

1. Explain the process of isolating DNA.
2. Define transformation and auxotroph.
3. Isolate DNA and use it to transform an auxotroph.

Background

Scientists from many different disciplines have used the technique of genetic engineering to genetically alter bacteria and eukaryotic organisms. In **genetic engineering,** genes are isolated from one organism and incorporated by manipulation in a laboratory into other bacterial or eukaryotic cells.

The first step in genetic engineering is extraction of DNA from the cell that possesses the desired gene. To extract DNA from bacteria, we disrupt the cells by chemical or mechanical means or by osmotic pressure.

The resuspension buffer used in this experiment contains lysozyme and RNase. What is the purpose of each of these enzymes? _____

_____

The 5 M NaCl denatures proteins such as DNase and stabilizes the DNA by forming an Na^+ shell around the negatively charged phosphates of the DNA. The DNA is precipitated as a viscous, translucent mass with cold alcohol.

Transformation is a rare event involving acquisition by a *recipient* bacterium of small pieces of DNA released from a dead *donor* bacterium. The acquired pieces of DNA can give the recipient bacterium new characteristics. In this experiment, you will extract DNA from an auxotrophic mutant strain of *Bacillus subtilis*. This strain, designated met⁻, cannot synthesize the amino acid methionine. You will use its DNA to transform a different *B. subtilis* mutant that cannot synthesize the amino acid phenylalanine (phe⁻). Neither of these bacteria can grow on media that lack the required amino acids. Glucose–minimal salts agar (GMSA) is such a medium (Table 28.1). GMSA contains glucose as a carbon and energy source; the other macronutrients, nitrogen, phosphate, and sulfur, are provided as inorganic salts. However, this medium does not provide amino acids, so only bacteria that can synthesize their own amino acids will be able to grow on it.

Materials

Centrifuge tube

Resuspension buffer

SDS

Proteinase

5 M NaCl

95% ethyl alcohol, ice cold

Glass rod

Graduated cylinder

Sterile test tube

Petri plates containing glucose–minimal salts agar (minimal medium) (3)

Spreading rod

Alcohol

1 ml sterile pipettes (2)

Pipette bulb or propipette

Water baths, 37°C and 60°C

Centrifuge

Cultures

Bacillus subtilis met⁻, phe⁺

B. subtilis met⁺, phe⁻

Techniques Required

Aseptic technique, Exercise 10

Spreading rod, Exercise 27

Pipetting, Appendix A

Procedure*

Isolation of DNA

1. Centrifuge the *Bacillus* met⁻, phe⁺ culture for 10 minutes at 3,000 rpm, to collect cells in the bottom of the tube.

*Adapted from Nick Kapp. Unpublished manuscript, San Bruno, CA: Skyline College, n.d.

Balance the centrifuge by placing a tube containing 10 ml water opposite your culture tube.

 Discard the supernatant (liquid portion) by carefully pouring it into a container of disinfectant.

2. Add 5 ml of resuspension buffer to the centrifuge tube and mix to resuspend the pellet. Incubate the tube in a 37°C water bath for 30 minutes.

3. Add 0.1 ml SDS and 50 μl proteinase. Why did the appearance of the suspension change? _____
_____

4. Incubate the tube in a 60°C water bath for 30 minutes. Remove the tube from the water bath and let it cool to room temperature.

5. Add 1 ml 5 M NaCl and mix gently.

6. Gently add 20 ml cold ethyl alcohol by letting it run down the side of the tube. You should have two distinct layers. Which layer (alcohol or water) is on top? _____

7. Let the tube stand in a test tube rack for 10 minutes. You should see a precipitate form at the interface of the two liquids. What is the precipitate? _____
_____

8. Gently mix the two layers with a glass rod using a circular motion and spool the fibers of DNA onto the rod. (See color plate II.2.) Describe the appearance of your spooled DNA.

9. Place the rod with the spooled DNA into a sterile test tube.

Transformation

1. Label three minimal media plates: "phe⁻ cells," "met⁻ DNA," and "DNA + phe⁻ cells."

2. Pipette 0.1 ml of the *Bacillus* met⁺, phe⁻ onto the appropriate two plates and spread with a sterile spreading rod (Figure 27.3).

⚠ **Dip the spreading rod in alcohol, then flame off the excess alcohol. After spreading, dip the spreading rod in alcohol and ignite it. Keep the flame away from the beaker of alcohol.**

3. Spread your spooled DNA over the surface of the appropriate two plates. Let the plates stand for about 15 minutes.

4. Incubate the plates at 37°C for 48 hours. Which plates are the controls? _____

5. Observe the plates and record your results.

exercise 29

Transformation of Bacteria

Name _____

Date _____

Lab Section _____

Purpose _____

_____

_____

Results

Describe the appearance, color, texture, and shape of your spooled DNA. _____

_____

	Inoculum		
	Bacillus phe⁻	met⁻ DNA	DNA + *Bacillus* phe⁻
Do you expect growth?			
Number of colonies			

Conclusions

Did you transform *Bacillus?* How do you know? _____

_____

_____

Questions

1. How were the *Bacillus* cells disrupted in this exercise? _____

_____

2. What physical characteristics of DNA allow it to be spooled onto a glass rod? Why is it not possible to spool out

precipitated proteins? _____

_____

_____

3. Why do you suspect that your spooled product is DNA and not RNA? _____

_____

_____

4. What new gene were you looking for in the transformed *Bacillus*? _____

_____

_____

Critical Thinking

1. If you got growth on the "DNA + *Bacillus* phe⁻" plate, how could you rule out contamination? Mutation?

2. Design an experiment to determine whether *Bacillus* DNA could be used to transform any other species.

3. It appears that *N. gonorrhoeae* acquired an antibiotic resistance gene (*tet*M) from *Streptococcus*. How did the *Neisseria* do this? How could you prove it was from *Streptococcus*?

exercise 30

DNA Fingerprinting

Objectives

After completing this exercise, you should be able to:

1. Define restriction enzyme and RFLP.
2. Perform agarose gel electrophoresis.
3. Identify your unknown.

Background

Restriction enzymes recognize specific short nucleotide sequences in double-stranded DNA and cleave both strands of the molecule (Table 30.1). Restriction enzymes are bacterial enzymes that restrict the host range of bacteriophages. The enzymes were discovered in laboratory experiments when phages were used to infect bacteria other than their usual hosts. Restriction enzymes in the new host destroyed the phage DNA. Today, over 300 restriction enzymes have been isolated and purified for use in DNA research. Each enzyme recognizes a different nucleotide sequence in DNA. The enzymes are named with three-letter abbreviations for the bacteria from which they were isolated.

DNA cleaved with restriction enzymes produces **restriction fragment length polymorphisms (RFLPs).** The size and number of the pieces is determined by **agarose gel electrophoresis** (Appendix G). The digested DNA is placed near one end of a thin slab of agarose and immersed in a buffer to allow current to flow through the agarose. Electrodes are attached to both ends of the gel, and a current is applied. Each piece of DNA then migrates toward the positive electrode at a rate determined by its size. The DNA fragments are visualized by staining with methylene blue or ethidium bromide.

These enzymes can be used to characterize DNA because a specific restriction enzyme will cut a molecule of DNA everywhere a specific base sequence occurs. When the DNA molecules from two different microorganisms are treated with the same restriction enzyme, the restriction fragments that are produced can be separated by electrophoresis. A comparison of the number and sizes of restriction fragments produced from different organisms provides information about their genetic similarities and differences—the more similar the patterns, the more closely related the organisms are expected to be.

In forensic science, analyses of DNA can be used to determine the father of a child or the perpetrator of a crime. DNA samples are digested to produce RFLPs that are separated by gel electrophoresis. The DNA fragments are transferred to a membrane filter by blotting so the fragments on the paper are in the same positions as the fragments on the gel. The paper is then heated to produce single strands of DNA. Small pieces of radioactively labeled single-stranded DNA called probes are used to distinguish core sequences of nucleotides. When the probe is added to the gel, it will anneal to its complementary strand. The gel can then be placed on photographic film, and the radioactive label will expose the film to produce an *autoradiograph*.

In this exercise, we will compare plasmids from an unknown bacterial species to plasmids from known bacteria in order to identify the unknown.

Materials
Digestion of DNA
DNA samples

Restriction enzyme

Table 30.1
Recognition Sequences of Some Restriction Endonucleases

Enzyme	Bacterial Source	Recognition Sequence
*Eco*R I	*Escherichia coli*	G↓AATTC CTTAA↑G
Hae III	*Haemophilus aegyptius*	GG↓CC CC↑GG

Restriction buffer

Micropipette, 1–10 μl

Micropipette tips (13)

Microcentrifuge tubes (5)

Electrophoresis of DNA Samples

Electrophoresis buffer

Casting tray and comb

Agarose, 0.8%

Tracking dye

Electrophoresis chamber and power supply

Ethidium bromide or methylene blue

Transilluminator or light box

Safety goggles

Techniques Required

Pipetting, Appendix A

Electrophoresis, Appendix G

Procedure

Every pair of students will perform five digests: One each from the unknown and four known bacteria.

Digestion of DNA

1. Label five microcentrifuge tubes "U," "1," "2," "3," and "4."
2. To each tube add 5 μl restriction buffer. Using a new pipette tip, add 4 μl restriction enzyme to each tube.
3. Add the DNA samples to each tube as follows. Use a different pipette tip for each sample. Why? _____

Tube	DNA
U	Unknown, 4 μl
1	Species 1, 4 μl
2	Species 2, 4 μl
3	Species 3, 4 μl
4	Species 4, 4 μl

4. Centrifuge the tubes for 1 to 2 seconds to mix. *Be sure to balance the centrifuge.* Then incubate the tubes in a 37°C water bath for 45 minutes. After in-

cubation, the tubes can be frozen if electrophoresis will be done during another lab period.

Electrophoresis of DNA Samples

1. Read about electrophoresis in Appendix G. Use the melted agarose to pour a gel, as described in Appendix G.
2. If necessary, defrost the tubes of digested DNA by holding them in your hand or placing them in a 37°C water bath.
3. Add 1 μl of tracking dye to each tube. Centrifuge the tubes for 1 to 2 seconds to mix. *Be sure the centrifuge is balanced.*
4. Load 14 μl of one of your samples prepared above into a well. Using a different pipette tip, load your other samples.
5. Once the wells have been filled, apply power (125V) to the chamber. During the run, you will see the tracking dye migrate. The tracking dye will separate into its two component dyes during electrophoresis. Turn off the power before the faster dye runs off the gel. Good separation of the DNA fragments occurs when the two dyes have separated by 4 or 5 cm. Run the gel until the dye is near the end of the gel, then turn the current off and remove the gel.
6. To stain the gel, use either step a or step b below. (See color plate II.3.)
    a. Transfer the gel to the ethidium bromide staining tray for 5 to 10 minutes.

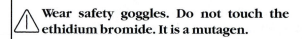

> ⚠ **Wear safety goggles. Do not touch the ethidium bromide. It is a mutagen.**

Transfer the gel to tap water to destain for 5 minutes. (Chlorine in tap water will inactivate residual ethidium.) Place your gel on the transilluminator and close the plastic lid.

> ⚠ **Do not look directly at the UV light. Do not turn the UV light on until the plastic lid is down.**

Ethidium bromide that is bound to DNA did not wash off and will fluoresce with UV light. Draw the bands or photograph your gel using the transilluminator camera.

    b. Transfer the gel to the methylene blue staining tray for 30 minutes to 2 hours. Destain by placing the gel in water for 30 minutes to overnight. Place your gel on a light box. Methylene blue that is bound to DNA did not wash off.
7. Carefully draw the location of the bands from each of the five digests.

exercise 30

DNA
Fingerprinting

Name _____

Date _____

Lab Section _____

Purpose _____
_____
_____

Data

Draw your gel so you have all five digests represented. Label the contents of each lane.

Conclusions

What is the identity of your unknown? How can you tell? _____
_____
_____

Questions

1. What restriction enzyme did you use? _____

   What nucleotide sequence does it recognize? _____

2. Using this map of pLAB30 (18 kb), give the number of restriction fragments and their lengths that would result from digesting pLAB30 with *Eco*R I, *Bam*H I, and both enzymes together.

Enzyme	Fragments Number	Size
*Eco*R I	_____	_____
*Bam*H I	_____	_____
*Eco*R I + *Bam*H I	_____	_____

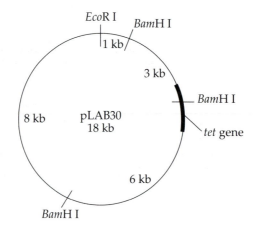

Which enzyme will give the smallest piece containing the tetracycline-resistance gene? _____

Critical Thinking

1. Differentiate between a gene and a RFLP.

2. How could you use this technique to trace the source of *E. coli* O157:H7?

   Why would you want to trace *E. coli*?

   Of what advantage is this technique over conventional biochemical tests?

3. The following agarose gel electrophoresis patterns were obtained from *Eco*R I digests of DNA from different isolates. The gels were developed with a DNA probe for a particular guanine-cytosine sequence. Which appear to be most closely related? Do any appear to be the same species? Briefly explain how you arrived at your conclusions.

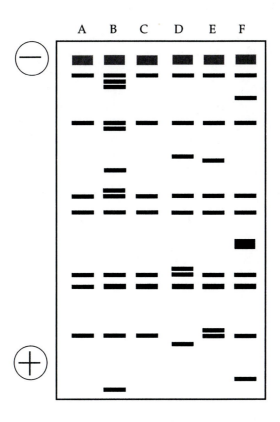

exercise 31

Genetic Engineering

Objectives

After completing this exercise, you should be able to:

1. Define transformation, restriction enzyme, and ligation.
2. Analyze DNA by electrophoresis.
3. Accomplish genetic change through transformation.

Background

For the genetic engineering of a bacterium using transformation, the desired gene is first inserted into a plasmid. Next, the recombinant plasmid is acquired by the bacterium. If the gene is expressed in the cell, the cell will produce a new protein product.

Genetic engineering is made possible by restriction enzymes. A **restriction enzyme** is an enzyme that recognizes and cuts only one particular sequence of nucleotide bases in DNA, and it cuts this sequence in the same way each time. These enzymes are divided into two major classes, based on their manner of cleavage. Class I enzymes bind to specific sequences but catalyze cleavage at sites up to thousands of base pairs from the binding sequence. Class II enzymes recognize specific sequences and catalyze cleavage within or next to these sequences. The recognition sequences usually include inverted repeats (*palindromes*) so that the same sequence is cut at the same point in both strands. Some of these enzymes catalyze cleavage of the two strands so that the ends of the resulting fragments are fully base-paired (**blunt ends**). Other enzymes catalyze a staggered cleavage of the two DNA strands so that resulting fragments have unpaired single strands (**sticky ends**) (Figure 31.1).

We will use the restriction enzymes *Hin*d III and *Bam*H I. *Hin*d III, an enzyme isolated from *Haemophi-lus influenzae*, cuts double-stranded DNA at the arrows in the sequence

$$A\downarrow AGCTT$$
$$TTCGA\uparrow A$$

*Bam*H I, isolated from *Bacillus amyloliquefaciens*, cuts double-stranded DNA at the arrows in the sequence

$$G\downarrow GATCC$$
$$CCTAG\uparrow G$$

In this exercise, we will use a plasmid containing an ampicillin-resistance gene (pAMP) and another plasmid with a kanamycin-resistance gene (pKAN) to transform an antibiotic-sensitive strain of *E. coli*. We will cut the two circular plasmids using restriction enzymes (Figure 31.2). The DNA fragments will be ligated using DNA ligase, and the ligated DNA will be used to transform the *E. coli*. *E. coli* does not normally take up new DNA; its cell wall will be damaged by treatment with $CaCl_2$ so DNA can enter. We will use agarose gel electrophoresis to determine whether the restriction enzymes cut the plasmids.

Materials
Plasmid Digestion
Sterile microcentrifuge tubes (2)

Micropipette, 1–10 μl

Sterile micropipette tips (4)

pAMP plasmid

pKAN plasmid

Restriction enzyme buffer

*Bam*H I and *Hin*d III enzymes

37°C water bath

C-C-G-**G-A-A-T-T-C**-C-C-C-A-C-**G-G-C-C**-G-A
G-G-C-**C-T-T-A-A-G**-G-G-G-T-G-**C-C-G-G**-C-T

(a) Before cleavage

    C-C-G-**G**   **A-A-T-T**-C-C-C-C-A-C-**G-G**   C-C-G-A
    G-G-C-**C-T-T-A-A**   G-G-G-G-T-G-**C-C**   **G-G**-C-T

(b) After cleavage

Figure 31.1
*Eco*R I recognizes G↑AATTC and leaves sticky ends. *Hae* III recognizes GG↑CC to produce blunt ends.

Figure 31.2
Restriction maps of plasmids showing locations of antibiotic-resistance markers. DNA is measured in the number of base pairs (bp).

Electrophoresis of Plasmids

Microcentrifuge tubes (3)

Micropipette, 1–10 μl

Sterile micropipette tips (7)

Casting tray and comb

Agarose, 0.8%

pAMP plasmid

Electrophoresis buffer

Tracking dye

Electrophoresis chamber and power supply

Ethidium bromide or methylene blue stain

Transilluminator or light box

Safety goggles

Ligation of DNA Pieces

Sterile microcentrifuge tubes (2)

Micropipette, 1–10 μl

Sterile micropipette tips (5)

Ligation buffer

65°C water bath

Ice

Sterile distilled water

DNA ligase

Transformation of Competent amp^s/kan^s *E. coli* Cells

Plate containing nutrient agar

Sterile microcentrifuge tubes (2)

Micropipettes, 1–10 μl and 200–1000 μl

Sterile micropipette tips (4)

50 mM CaCl$_2$

Ice

43°C water bath

35°C water bath

Sterile nutrient broth

Culturing Transformed Cells

Micropipette, 10–100 μl

Sterile micropipette tips (2)

Plates containing nutrient agar (3)

Plates containing ampicillin–nutrient agar (3)

Plates containing kanamycin–nutrient agar (3)

Plates containing ampicillin and kanamycin–nutrient agar (2)

Spreading rod and alcohol

Culture

E. coli amp^s/kan^s

Techniques Required

Inoculating loop, Exercise 10

Spreading rod, Exercise 27

Pipetting, Appendix A

Electrophoresis, Appendix G

Procedure (Figure 31.3)

A. Plasmid Digestion
Keep all reagents ice cold.

1. Label two sterile microcentrifuge tubes "1" and "2."
2. Aseptically add reagents as shown below.

Tube	pAMP	pKAN	Restriction Buffer	*Bam*H I/ *Hin*d III
1	5.5 μl	—	7.5 μl	2 μl
2	—	5.5 μl	7.5 μl	2 μl

3. Close the caps and mix the tubes for 1 to 2 seconds by centrifuging.

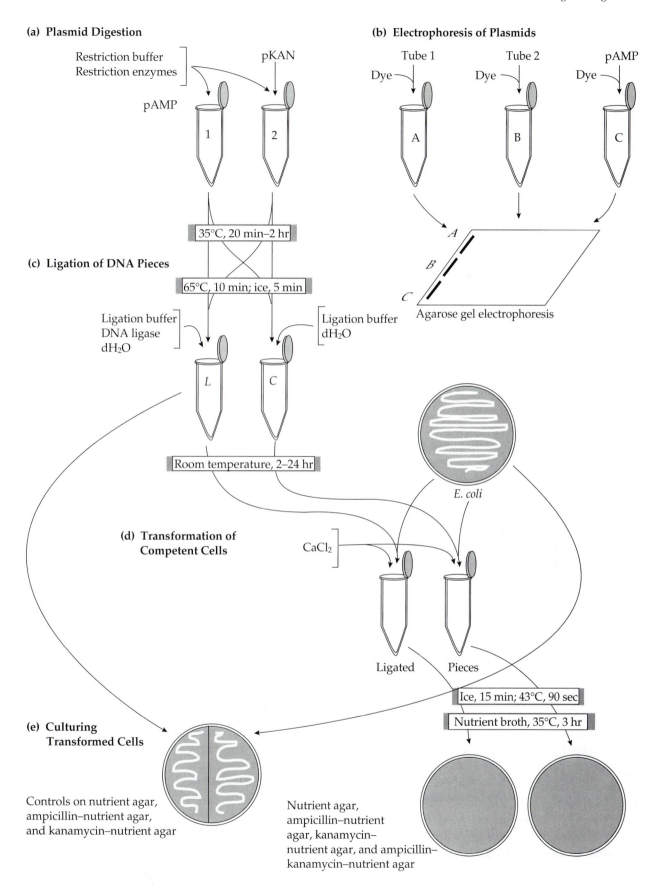

(a) Plasmid Digestion

Restriction buffer
Restriction enzymes

pAMP

pKAN

1

2

(b) Electrophoresis of Plasmids

Tube 1

Tube 2

pAMP

Dye

Dye

Dye

A

B

C

35°C, 20 min–2 hr

A

B

C

Agarose gel electrophoresis

(c) Ligation of DNA Pieces

65°C, 10 min; ice, 5 min

Ligation buffer
DNA ligase
dH₂O

Ligation buffer
dH₂O

L

C

Room temperature, 2–24 hr

E. coli

(d) Transformation of Competent Cells

CaCl₂

Ligated

Pieces

Ice, 15 min; 43°C, 90 sec

Nutrient broth, 35°C, 3 hr

(e) Culturing Transformed Cells

Controls on nutrient agar,
ampicillin–nutrient agar,
and kanamycin–nutrient agar

Nutrient agar,
ampicillin–nutrient
agar, kanamycin–
nutrient agar, and ampicillin–
kanamycin–nutrient agar

Figure 31.3

Procedure for plasmid digestion, electrophoresis, ligation, and transformation of *E. coli*.

> ⚠ **Be sure the centrifuge is balanced, by placing your tubes opposite each other.**

4. Incubate the tubes at 35°C for 20 minutes to 2 hours, as instructed. Tubes can be frozen until the next lab period.

B. Electrophoresis of Plasmids

1. Read about electrophoresis in Appendix G.
2. Use the melted agarose to pour a gel, as described in Appendix G.
3. Prepare three tubes as listed below. Remove samples aseptically from tubes 1 and 2. Freeze tubes 1 and 2 for procedure part C.

Tube	Uncut pAMP	Sample from Part A Tube #1	Sample from Part A Tube #2	Tracking Dye
A	—	5 µl	—	1 µl
B	—	—	5 µl	1 µl
C	5 µl	—	—	1 µl

4. Mix the tubes by centrifuging for 1 to 2 seconds.
5. Fill the electrophoresis chamber with electrophoresis buffer.
6. If necessary, defrost the tubes prepared in part A by holding them in your hand or placing them in a 37°C water bath. Load 5 µl from tube A into the first well of your gel. Load 5 µl from tube B into the second well and 5 µl from tube C into the third well.
7. Place the gel in the electrophoresis chamber, attach the lid, and apply 125V to the chamber. During the run, you will see the tracking dye migrate. The tracking dye migrates faster than the DNA. The dye is added so you can turn the power off before the DNA runs off the gel. Why do the two dyes separate? _____ Turn off the current when the two dyes have separated by 4 or 5 cm.
8. To stain the gel, use either step a or step b below. (See color plate II.3.)
   a. Transfer the gel to the ethidium bromide staining tray for 5 to 10 minutes.

> ⚠ **Wear safety goggles. Do not touch the ethidium bromide. It is a mutagen.**

Transfer the gel to tap water to destain for 5 minutes. (Chlorine in tap water will inactivate residual ethidium.) Place your gel on the transilluminator and close the plastic lid.

> ⚠ **Do not look directly at the UV light. Do not turn the UV light on until the plastic lid is down.**

Ethidium bromide that is bound to DNA did not wash off and will fluoresce with UV light. Draw the bands or photograph your gel using the transilluminator camera.
   b. Transfer the gel to the methylene blue staining tray for 30 minutes to 2 hours. Destain by placing the gel in water for 30 minutes to overnight. Place your gel on a light box. Methylene blue that is bound to DNA did not wash off. Draw the bands.

C. Ligation of DNA Pieces

1. Label two sterile microcentrifuge tubes "L" for "ligated" to describe the DNA and "C" for control (unligated).
2. Heat tubes 1 and 2 from part A at 65°C for 10 minutes to destroy the restriction enzymes. Why is this necessary? _____
3. Cool tubes 1 and 2 on ice for 5 minutes. Why is cooling necessary? _____
4. Aseptically add reagents to tubes L and C, as shown below.

Tube	Ligation Buffer	Sterile dH₂O	Tube 1	Tube 2	DNA Ligase
L	10 µl	3 µl	3 µl	3 µl	1 µl
C	10 µl	4 µl	3 µl	3 µl	—

5. Close caps securely and centrifuge the tubes for 1 to 2 seconds.
6. Incubate the tubes at room temperature for 2 to 24 hours.

D. Transformation of Competent amp^s/kan^s *E. coli* Cells

1. Label two sterile microcentrifuge tubes "Ligated" and "Pieces." Aseptically pipette 250 µl of ice-cold CaCl₂ into each tube. Keep the tubes on ice.
2. With a sterile loop, transfer a loopful of *E. coli* to each tube. Mix to make a suspension.
3. Aseptically add 10 µl from tube L to the tube labeled "Ligated." Save tube L for part E. Add 10 µl from tube C to the "Pieces" tube.
4. Incubate the tubes on ice for 15 minutes to allow the DNA to settle on the cells.
5. *Heat-shock* the cells by placing them in a 43°C water bath for 90 seconds so the DNA will enter the cells.

6. Return the tubes to the ice for 1 minute, then add 700 μl sterile nutrient broth. Incubate the tubes in a 35°C water bath for at least 3 hours.

E. Culturing Transformed Cells

1. Prepare the following controls. Inoculate one-half of a nutrient agar plate with a loopful of the amp^s/kan^s *E. coli* cells. Inoculate the other half with a loopful of the ligated plasmid DNA preparation (tube L). Similarly, inoculate an ampicillin–nutrient agar plate and a kanamycin–nutrient agar plate. How many plates do you have? _____

2. Add 100 μl from the "Ligated" tube to each of the following plates:

   Nutrient agar
   Ampicillin–nutrient agar
   Kanamycin–nutrient agar
   Ampicillin–kanamycin–nutrient agar

3. Sterilize a spreading rod by dipping it in alcohol, quickly igniting the alcohol, and letting the alcohol burn off.

 While it is burning, hold the spreading rod pointed *down*. Keep the beaker of alcohol away from the flame.

Let the spreading rod cool.

4. Spread cells over one plate with the sterile spreading rod. Sterilize the spreading rod before spreading the cells over each of the remaining plates.

5. Repeat steps 2 through 4 to inoculate four plates from the "Pieces" tube.

6. Label the plates and incubate them for 24 hours at 35°C.

7. Observe the plates and record your results.

exercise 31

Genetic Engineering

Name _____

Date _____

Lab Section _____

Purpose _____

_____

_____

Data

Sketch your electrophoresis results. Note the approximate sizes of the DNA fragments.

Controls:

Inoculum	Growth on		
	Nutrient Agar	Ampicillin-Nutrient Agar	Kanamycin-Nutrient Agar
amp^s/kan^s *E. coli*			
DNA			

Transformation:

Inoculum	Number of Colonies			
	Nutrient Agar	Ampicillin-Nutrient Agar	Kanamycin-Nutrient Agar	Ampicillin-Kanamycin-Nutrient Agar
Ligated				
Pieces				

Conclusions

1. Did you cut the plasmids? How do you know? _____

    _____

2. What do the results of your transformation experiment indicate? _____

    _____

    _____

Questions

1. What is the purpose of each control? _____

    _____

    _____

    _____

2. If there was no growth on the DNA+$E.$ $coli$ plates, what went wrong? _____

    _____

    _____

3. Why were the $E.$ $coli$ bacteria treated with $CaCl_2$ before the plasmids were added? _____

    _____

    _____

4. Do the enzymes used in this experiment produce blunt or sticky ends? Why was this type of enzyme used? ____

    _____

    _____

Critical Thinking

1. If you got growth on the DNA + $E.$ $coli$ plates, how could you rule out contamination? Mutation?

2. How could you prove ligation of the two plasmids occurred and not two separate transformation events?

3. Using the following gel, determine whether recombination between the plasmid and chromosome occurred. Briefly explain how you can tell it is transformation.

Lane A: *Eco*R I digest of cell's DNA.
Lane B: Undigested plasmid.
Lane C: *Eco*R I digest of plasmid.
Lane D: *Eco*R I digest of cell's DNA after a transformation experiment.
Lane E: *Hin*d III digest of cell's DNA.
Lane F: *Hin*d III digest of plasmid.
Lane G: *Hin*d III digest of cell's DNA after a transformation experiment.

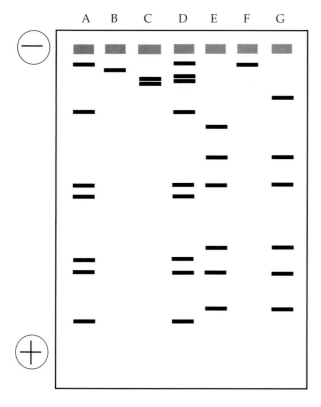

exercise 32

Ames Test for Detecting Possible Chemical Carcinogens

The requirements of health can be stated simply. Those fortunate enough to be born free of significant congenital disease or disability will remain well if three basic needs are met: they must be adequately fed; they must be protected from a wide range of hazards in the environment; and they must not depart radically from the pattern of personal behavior under which man evolved, for example, by smoking, overeating, or sedentary living.

THOMAS McKEOWN

Objectives

After completing this exercise, you should be able to:

1. Differentiate between the terms mutagenic and carcinogenic.
2. Provide the rationale for the Ames test.
3. Perform the Ames test.
4. Differentiate between a revertant and a mutant.

Background

Every day we are exposed to a variety of chemicals, some of which are **carcinogens**—that is, they can induce cancer. Historically, animal models have been used to evaluate the carcinogenic potential of a chemical, but the procedures are costly and time-consuming and result in the inadequate testing of some chemicals. Many chemical carcinogens induce cancer because they are **mutagens** that alter the nucleotide base sequence of DNA.

Bruce Ames and his coworkers at the University of California at Berkeley have developed a fast, inexpensive assay for mutagenesis using a *Salmonella* auxotroph. Most chemicals that have been shown to cause cancer in animals have proven mutagenic in the **Ames test.** Since not all mutagens induce cancer, the Ames test is a screening technique that can be used to identify high-risk compounds that must then be tested for carcinogenic potential. The Ames test uses auxotrophic strains of *Salmonella typhimurium,* which cannot synthesize the amino acid histidine (his⁻). The strains are also defective in *dark excision* repair of mu-

tations (***uvrB***), and an ***rfa*** mutation eliminates a portion of the lipopolysaccharide that coats the bacterial surface. The *rfa* mutation prevents the *Salmonella* from growing in the presence of sodium desoxycholate or crystal violet and increases the cell wall permeability; consequently, more mutagens enter the cell. The *uvrB* mutation minimizes repair of mutations; as a result, the bacteria are much more sensitive to mutations. To grow the auxotroph, histidine and biotin (due to the *uvrB*) must be added to the culture media.

In the Ames test, a small sample of the test chemical (a suspected mutagen) is either added to a melted soft agar (containing histidine) suspension of the *Salmonella* that is overlayed on glucose–minimal salts agar (GMSA) (lacking histidine) or is placed on the surface of GMSA seeded with a lawn of the *Salmonella* auxotroph. (A small amount of histidine allows all the cells to go through a few divisions.) If certain mutations occur, the bacteria may *revert* to a wild type, or prototroph, and grow to form a colony. Only bacteria that have mutated (reverted) to his⁺ (able to synthesize histidine) will grow into colonies. In theory, the number of colonies is proportional to the mutagenicity of the chemical. (See color plate II.1.)

In nature many chemicals are neither carcinogenic nor mutagenic, but they are metabolically converted to mutagens by liver enzymes. The original Ames test could not detect the mutagenic potential of these in vivo (in the body) conversions. Suspect chemicals can be incubated with a suspension of rat liver enzyme preparation in an oxygenated environment to be "activated." *Salmonella* is exposed to the "activated" chemicals to test for mutagenicity (Figure 32.1a). In this

Figure 32.1

The Ames test. **(a)** Rat liver extract can be used to activate a suspected mutagen. **(b)** A disk impregnated with a suspected mutagen is placed on the agar. The agar lacking histidine was inoculated with mutant *Salmonella* unable to synthesize histidine (his−). **(c)** Only bacteria that have further mutated (reverted) to his+ (able to synthesize histidine) will grow into colonies.

experiment, the chemicals to be evaluated are mutagenic and will not require activation (Figure 32.1b).

> ☣ ***Salmonella typhimurium* is a potential pathogen.**

Materials

Petri plates containing glucose–minimal salts agar (2)

Tubes containing 4 ml soft agar (glucose–minimal salts + 0.05 mM histidine and 0.05 mM biotin) (2)

Sterile filter paper disks

Sterile 1 ml pipette

Forceps and alcohol

Paper disks soaked with suspected mutagens:
    Benzo (α) pyrene
    Ethidium bromide
    Nitrosamine
    2-aminofluorene
    Cigarette smoke condensates
    Food coloring
    Hair dye
    Household compound (your choice)
    Maraschino cherries
    Hot dog
    Cosmetic or personal care item (your choice)

Culture

Salmonella typhimurium his−, *uvrB, rfa*

Techniques Required

Pipetting, Appendix A

Aseptic technique, Exercise 10

Procedure*

1. Label one glucose–minimal salts agar plate "Control" and the other "Test."
2. Aseptically pipette 0.1 ml *Salmonella* to one of the soft agar tubes, and quickly pour it over the surface of the control plate. Tilt the plate back and forth gently to spread the agar evenly. Let it harden.
3. Obtain another soft agar tube and repeat step 2 for the test plate. Why does the soft agar contain histidine? _____
   _____

> ⚠ **Dip the forceps in alcohol, and with the tip pointed down, burn off the alcohol. Keep the beaker of alcohol away from the flame.**

*Adapted from D. J. Pringnitz. "The Effects of Ascorbic Acid on the Ability of Dimethylnitrosamine to Cause Reversion in a *Salmonella typhimurium* Mutant." Master's thesis. Mankato, MN: Mankato State University, 1975.

4. Aseptically place three to five disks saturated with the various suspected mutagens on the surface of the soft agar overlay of the test plate. If the chemical is crystalline, place a few crystals directly on the test plate. Label the bottom of the Petri plate with the name of each chemical. Place three sterile disks on the control plate.

⚠️ ***Be very careful.*** **These are potentially dangerous compounds.**

5. Incubate both plates, right-side up, at 35°C for 48 hours.
6. Describe your results. Look for revertant colonies surrounding the disk as shown in color plate II.1.

exercise 32

Ames Test for Detecting Possible Chemical Carcinogens

Name _____

Date _____

Lab Section _____

Purpose _____

_____

_____

Data

Show the location of paper disks and bacterial growth on the plates. Number the disks to correspond to the table below.

Control plate

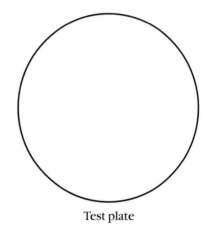

Test plate

Complete the following table.

Compound	Growth? (large colonies)	Mutagenic?
1		
2		
3		
4		
5		

Questions

1. What degree of spontaneous reversion occurred? _____

_____

_____

2. Why are mutants used as test organisms in the Ames test? _____

_____

_____

3. What is the advantage of this test over animal tests? Disadvantages? _____

_____

_____

_____

Critical Thinking

1. Does this technique give a minimum or maximum mutagenic potential? Briefly explain.

2. Is it possible for a chemical to be negative in the Ames test yet be a carcinogen? Briefly explain.

3. The Ames test was used to determine the mutagenic capability of 2-aminofluorene. What can you conclude from these data?

Test Substance	Relative Amounts of Growth in the Ames Test
2-Aminofluorene	+
2-Aminofluorene activated by:	
Liver enzymes	+ + + +
Intestinal enzymes	+ +
Bacteroides fragilis enzymes	+ +
Intestinal enzymes + *B. fragilis* enzymes	+ + +

part eight
The Microbial World

EXERCISES

Organisms with eukaryotic cells include algae, protozoa, fungi, and higher plants and animals. The eukaryotic cell is typically larger and structurally more complex than the prokaryotic cell. The DNA of a eukaryotic cell is enclosed within a membrane-bounded **nucleus.** In addition, eukaryotic cells contain membrane-bounded **organelles,** which have specialized structures and perform specific functions (see the illustration on the next page).

Yeasts are possibly the best known microorganisms. They are widely used in commercial processes and can be purchased in the supermarket for baking. Yeasts are unicellular fungi (Exercise 34).

Van Leeuwenhoek was the first to observe the yeast responsible for fermentation in beer:

> *I have made divers observations of the yeast from which beer is made and I have generally seen that it is composed of globules floating in a clear medium (which I judged to be the beer itself).* *

Many algae are only visible through the microscope, while others can be a few meters long. Algae are important producers of oxygen and food for

*Quoted in H. A. Lechevalier and M. Solotorovsky. *Three Centuries of Microbiology*. New York: Dover Publications, 1974, p. 502.

protozoa and other organisms. A few unicellular algae, such as the agents of "red tides," *Alexandrium catanella* and related species, are toxic to animals, including humans, when ingested in large numbers.

Algae and cyanobacteria (Exercise 36) are photoautotrophs; fungi and protozoa are all chemoheterotrophs. Protozoa, originally called "infusoria," were of interest to early investigators. In 1778, Friedrich von Gleichen studied food vacuoles by feeding red dye to his infusoria. A refinement of this experiment will be done in Exercise 37. The compound light microscope is invaluable for identification of molds, algae, cyanobacteria, and protozoa; however, many yeasts and bacteria look alike through the microscope. These organisms are identified by biochemical testing.

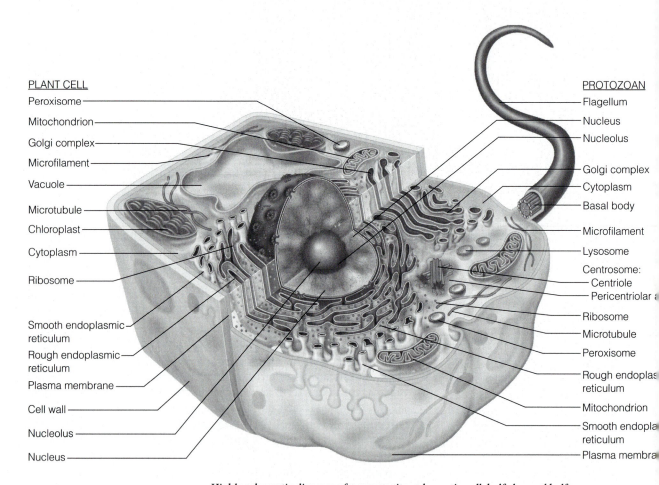

PLANT CELL
- Peroxisome
- Mitochondrion
- Golgi complex
- Microfilament
- Vacuole
- Microtubule
- Chloroplast
- Cytoplasm
- Ribosome
- Smooth endoplasmic reticulum
- Rough endoplasmic reticulum
- Plasma membrane
- Cell wall
- Nucleolus
- Nucleus

PROTOZOAN
- Flagellum
- Nucleus
- Nucleolus
- Golgi complex
- Cytoplasm
- Basal body
- Microfilament
- Lysosome
- Centrosome: Centriole
- Pericentriolar a
- Ribosome
- Microtubule
- Peroxisome
- Rough endoplas reticulum
- Mitochondrion
- Smooth endopla reticulum
- Plasma membra

Highly schematic diagram of a composite eukaryotic cell, half alga and half protozoan.

exercise 33

Unknown Identification and Bergey's Manual

The strategy of discovery lies in determining the sequence of choice of problems to solve. Now it is in fact very much more difficult to see a problem than to find a solution to it. The former requires imagination, the latter only ingenuity.

JOHN BERNAL

Objectives

After completing this exercise, you should be able to:

1. Explain how bacteria are characterized and classified.
2. Use *Bergey's Manual.*
3. Identify an unknown bacterium.

Background

In microbiology, a system of classification must be available to allow the microbiologist to categorize and classify organisms. Communication among scientists would be very limited if no universal system of classification existed. Until recently, the **taxonomy** (grouping) of bacteria was difficult because few definite anatomical or visual differences exist. With these limitations, most bacteria are identified through evaluation of primary characteristics, such as morphology and growth patterns, and secondary characteristics, such as metabolism and serology. A characteristic that is critical for distinguishing one bacterial group from another may be irrelevant for characterization of other bacteria.

The most important reference for bacterial taxonomy is **Bergey's Manual,** * in which bacteria are classified according to similarities in their ribosomal RNA (rRNA). The Domains Bacteria and Archaea are included in *Bergey's Manual.* Bacteria and Archaea with similar rRNA sequences are grouped together into taxa. The taxa used for the Bacteria and Archaea are phylum, class, order, family, genus, and species. Although the characteristics of a given group are relatively constant, through repeated laboratory culture, atypical bacteria

will be found. This variability, however, only heightens the fun of classifying bacteria.

You will be given an unknown heterotrophic bacterium to characterize and identify. By using careful deduction and systematically compiling and analyzing data, you should be able to identify the bacterium.

Keys to species of bacteria are provided in Appendix H, Figures H.1 and H.2. The key is an example of a dichotomous classification system—that is, a population is repeatedly divided into two parts until a description identifies a single member. Such a key is sometimes called an "artificial key" because there is no single correct way to write one. You may want to check your conclusion with the species description given in *Bergey's Manual.*

To begin your identification, ascertain the purity of the culture you have been given and prepare stock and working cultures. Avoid contamination of your unknown. Note growth characteristics and Gram stain appearance for clues about how to proceed. After culturing and staining the unknown, many bacterial groups can be eliminated. Final determination of your unknown will depend on careful selection of the relevant biochemical tests and weighing the value of one test over another in case of contradictions. (See color plate XII.) Enjoy!

Materials

Petri plates containing trypticase soy agar (2)

Trypticase soy agar slant (2)

All stains, reagents, and media previously used

Culture

Unknown bacterium # _____

---

*Bergey's Manual of Systematic Bacteriology, 2nd ed., 5 vols, (2000) is the reference for classification. *Bergey's Manual of Determinative Bacteriology,* 9th ed. (1994) is used for laboratory identification.

Techniques Required

Compound light microscopy, Exercise 1

Hanging drop, Exercise 2

Wet mount, Exercise 2

Negative staining, Exercise 4

Gram staining, Exercise 5

Acid-fast staining, Exercise 6

Endospore, capsule, and flagella staining, Exercise 7

Inoculating loop and needle, Exercise 10

Aseptic technique, Exercise 10

Plate streaking, Exercise 11

OF test, Exercise 13

Starch hydrolysis, Exercise 13

MRVP tests, Exercise 14

Fermentation tests, Exercise 14

Protein catabolism, Parts 1 and 2, Exercises 15 and 16

Catalase test, Exercise 17

Nitrate reduction test, Exercise 17

Oxidase test, Exercise 17

Procedure

1. Streak your unknown onto the agar plates for iso-
    lation. Incubate one plate at 35°C and the other at
    room temperature for 24 to 48 hours. Note the
    growth characteristics and the temperature at
    which each one grew best.

2. Aseptically inoculate two trypticase soy agar slants.
    Incubate them for 24 hours. Describe the resulting
    growth. Keep both slant cultures in the refrigera-
    tor. One is your stock culture; the other is your
    working culture. Subculture onto another slant
    from your stock culture when your working cul-
    ture is contaminated or not viable. Keep the work-
    ing culture in the refrigerator when it is not in use.

3. Use your working culture for all identification pro-
    cedures. When a new slant is made and its purity
    demonstrated, discard the old working culture.
    What should you do if you think your culture is
    contaminated? _____
    _____

4. Read the keys in Appendix H, Figures H.1 and H.2,
    to develop ideas on how to proceed. Perhaps de-
    termining staining characteristics might be a good
    place to start. What shape is it? _____
    What can be eliminated? _____

5. After determining its staining and morphologic
    characteristics, determine which biochemical tests
    you will need. Do not be wasteful. Inoculate *only*
    what is needed. It is not necessary to repeat a
    test—do it once accurately. Do not perform un-
    necessary tests.

6. If you come across a new test—one not previously
    done in the course—determine if it is essential.
    Can you circumvent it? _____
    If not, consult your instructor.

exercise 33

Unknown Identification and Bergey's Manual

Name _____

Date _____

Lab Section _____

Purpose _____

_____

Data

Write "not tested" (NT) next to tests that were not performed. Unknown # _____

Morphological, Staining, and Cultural Characteristics	Sketches: Label and give magnification (_____ ×)			
The cell				
Staining characteristics				
Gram _____ Age _____				
Other _____ Age _____				
Shape _____				
Size _____				
Arrangement _____				
Endospores (position) _____				
_____				
Motility _____	**Results: Essential biochemical characteristics**			
Determined by _____		Time (hr)		
Colonies on Petri plate		24	48	. .
Diameter _____					Temp. ____°C
Appearance _____	Glucose			
Color _____	Lactose			
Elevation _____	Mannitol			
Margin _____	Catalase			
Consistency _____	Oxidase			
Agar slant Age _____	H₂S			
Amount of growth _____	Nitrate reduction			
Pattern _____	Indole			
Color _____	Methyl red			
Growth in broth Age _____	V–P			
Amount of growth _____	Citrate			
Pellicle _____				
Flocculant _____				
Sediment _____				

Motility _____

Determined by _____

Results: Essential biochemical characteristics

	Time (hr)		
	24	48	. .
Glucose			
Lactose			
Mannitol			
Catalase			
Oxidase			
H₂S			
Nitrate reduction			
Indole			
Methyl red			
V–P			
Citrate			

Temp. ____°C

Abbreviations:
A = Acid
G = Gas
a = slight acid
alk = alkaline
+ = positive
− = negative
ng = no growth

Other characteristics, special media, etc. _____

_____

_____

Conclusion

What organism was in unknown # _____ ? _____

Questions

1. On a separate sheet of paper, write your rationale for arriving at your conclusion.

2. Why is it necessary to complete the identification of a bacterium based on its physiology rather than its

   morphology? _____

   _____

   _____

   _____

3. Place the following organisms next to their description in the key shown below:

Actinomyces	*Escherichia*	*Pseudomonas*
Bacillus	*Lactobacillus*	Purple sulfur bacteria
Bacteroides	*Mycobacterium*	*Staphylococcus*
Campylobacter	*Mycoplasma*	*Streptococcus*
Clostridium	*Neisseria*	*Streptomyces*
Corynebacterium	*Nitrosomonas*	*Thiobacillus*
Cyanobacteria	*Nocardia*	*Treponema*

   I. Gram-positive

      A. Rods

         1. Endospore-forming

            a. Obligate anaerobe   _____

            b. Facultative anaerobe   _____

         2. Non-endospore-forming

            a. Acid-fast

               (1) Mycelia formed   _____

               (2) Mycelia not formed   _____

            b. Not acid-fast

               (1) Regular   _____

               (2) Club-shaped; stain irregularly   _____

         3. Produce conidia

            a. Aerobic   _____

            b. Anaerobic   _____

      B. Cocci

         1. Oxidative; catalase-positive   _____

         2. Fermentative; catalase-negative   _____

      C. Wall-less   _____

II.   Gram-negative

    A.   Heterotrophic

        1.   Helical or vibroid

            a.   Possess axial filament   _____

            b.   Possess flagella   _____

        2.   Rods

            a.   Aerobic   _____

            b.   Facultatively anaerobic   _____

            c.   Anaerobic   _____

        3.   Cocci   _____

    B.   Autotrophic

        1.   Photoautotrophic

            a.   Produce O_2 _____

            b.   Do not produce O_2 _____

        2.   Chemoautotrophic

            a.   Oxidize N-containing inorganic compounds   _____

            b.   Oxidize S-containing inorganic compounds   _____

exercise 34

Fungi: Yeasts

Objectives

After completing this exercise, you should be able to:

1. Culture and identify yeasts.
2. Differentiate between yeasts and bacteria.

Background

Fungi possess eukaryotic cells and can exist as unicellular or multicellular organisms. They are heterotrophic and obtain nutrients by absorbing dissolved organic material through their cell walls and plasma membranes. Fungi (with the exception of yeasts) are aerobic. Unicellular yeasts, multicellular molds, and macroscopic species such as mushrooms are included in the Kingdom Fungi. Fungi generally prefer more acidic conditions and tolerate higher osmotic pressure and lower moisture than bacteria. They are larger than bacteria, with more cellular and morphologic detail. In contrast to bacterial characterization, primary characteristics, such as morphology and cellular detail, are used to classify fungi, with little attention given to secondary characteristics, such as metabolism and antigenic composition. Fungi are structurally more complex than bacteria but are less diverse metabolically.

Yeasts are nonfilamentous, unicellular fungi that are typically spherical or oval in shape. Yeasts are widely distributed in nature, frequently found on fruits and leaves as a white, powdery coating. Yeasts reproduce asexually by budding, a process in which a new cell forms as a protuberance (bud) from the parent cell (Figure 34.1a). In some instances, when buds fail to detach themselves, a short chain of cells called a **pseudohypha** forms (Figure 34.1b). When yeasts reproduce sexually, they may produce one of several types of sexual spores. The type of sexual spore produced by a species of yeast is used to classify the yeast as to division (phylum). Metabolic activities are also used to identify genera of yeasts.

Yeasts are facultative anaerobes. Their metabolic activities are used in many industrial fermentation processes. Yeasts are used to prepare many foods, including bread, and beverages, such as wine and beer.

In the laboratory, **Sabouraud agar,** a selective medium, is commonly used to isolate yeast. Sabouraud agar has very simple nutrients (glucose and peptone) and a low pH, which inhibits the growth of most other organisms. Many of the techniques useful in working with bacteria can be applied to yeasts.

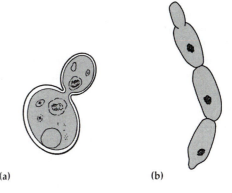

Figure 34.1
Budding yeasts. **(a)** A bud forming from a parent cell.
(b) Pseudohyphae are short chains of cells formed by some yeasts.

Materials

Glucose fermentation tubes (2)

Sucrose fermentation tubes (2)

Petri plates containing Sabouraud agar (2)

Bottle containing glucose–yeast extract broth

Sterile cotton swab

Coverslip

Test tube

Methylene blue (second period)

Balloon

Fruit or leaves

Cultures (as assigned)

Baker's yeast

Rhodotorula rubra

Candida albicans

Saccharomyces cerevisiae

Techniques Required

Compound light microscopy, Exercise 1

Colony morphology, Exercise 9

Wet mount, Exercise 2

Plate streaking, Exercise 11

Fermentation tests, Exercise 14

Procedure

Yeasts

1. Gently suspend a pinch of baker's yeast in a small amount of lukewarm water in a test tube, creating a milky solution.
2. Each pair of students will use one of the yeast cultures and the suspension of baker's yeast.
   a. Divide one Sabouraud agar plate in half. Streak one half with a known yeast culture and the other half with the baker's yeast suspension.
   b. Inoculate each organism into a glucose fermentation tube and a sucrose fermentation tube.
3. Incubate all media at 35°C until growth is seen.
4. Make a wet mount of each culture by using a small drop of methylene blue. Record your observations.
5. After the yeasts have grown, record your results. Examine cultures of the yeasts you did not

culture, and record pertinent results. (See color plate IV.1.)

Yeast Isolation

1. Cut the fruit or leaves into small pieces. Place them in the bottle of glucose–yeast extract broth. Cover the mouth of the bottle with a balloon. Incubate the bottle at room temperature until growth has occurred. Record the appearance of the broth after incubation. Was gas produced? _____
2. Divide a Sabouraud agar plate in half. Each partner inoculates half of the medium, following either procedure a or procedure b.
   a. Swab the surface of your tongue with a sterile swab. Inoculate one-half of the agar surface with the swab. Discard the swab in the disinfectant. Why will few bacteria grow on this medium? _____
   b. Using a sterile inoculating loop, streak one-half of the agar surface with a loopful of broth from the bottle just prepared in step 1.
3. Incubate the plate, inverted, at room temperature until growth has occurred. Prepare wet mounts with methylene blue from different-appearing colonies. Record your results.

exercise 34

Fungi: Yeasts

Name _____

Date _____

Lab Section _____

Purpose _____

_____

Data

Yeasts

Fermentation tubes:

Organism	Glucose			Sucrose		
	Acid	Gas	Fermentation	Acid	Gas	Fermentation
Rhodotorula rubra						
Candida albicans						
Saccharomyces cerevisiae						
Baker's yeast						

Sabouraud agar plates:

Organism	Draw a Typical Colony	Wet Mounts
Rhodotorula rubra Color: _____		
Candida albicans Color: _____		

Organism	Draw a Typical Colony	Wet Mounts
Saccharomyces cerevisiae Color: _____		
Baker's yeast Color: _____		

Yeast Isolation

Plants used: _____

Describe the appearance of the glucose broth after _____ days' incubation. _____

_____

Was gas produced? _____

Sabouraud agar:

Area sampled? _____

Colony Appearance	Color	Size	Wet Mounts

Any bacteria seen? _____

If so, which colonies? _____

Questions

1. Could you identify the genus of baker's yeast? _____

   _____

2. Did you culture yeast from your mouth? _____ From the plants? _____ How do you know? _____

   _____

3. What was the purpose of the balloon on the glucose–yeast extract broth bottle? _____

   _____

4. Define the term *yeast.* _____

   _____

   _____

   _____

   _____

5. Compare and contrast yeast and bacteria regarding their appearance both on solid media and under the

   microscope. _____

   _____

   _____

   _____

6. Why are yeast colonies larger than bacterial colonies? _____

   _____

   _____

   _____

   _____

Critical Thinking

A 45-year-old HIV-positive woman using continuous intravenous antibiotic infusion to manage a kidney infection developed a 38°C fever. Cultures of blood, the needle tip, and the insertion site revealed large ovoid cells that reproduced by budding.

1. What is your preliminary identification of the infecting organism?

2. How might the antibiotic treatment have affected this patient?

3. What do the findings suggest about the portal of entry for the infection?

exercise 35

Fungi: Molds

*And what is weed? A plant whose virtues
have not been discovered.*

RALPH WALDO EMERSON

Objectives

After completing this exercise, you should be able to:

1. Characterize and classify fungi.
2. Compare and contrast fungi and bacteria.
3. Identify common saprophytic molds.
4. Explain dimorphism.

Background

The multicellular filamentous fungi are called **molds.** Because of the wide diversity in mold morphology, morphology is very useful in classifying these fungi.

A macroscopic mold colony is called a **thallus** and is composed of a mass of strands called **mycelia.** Each strand is a **hypha,** with the **vegetative hyphae** growing in or on the surface of the growth medium. Aerial hyphae, called **reproductive hyphae,** originate from the vegetative hyphae and produce a variety of asexual reproductive **spores** (Figure 35.1). The hyphal strand of most molds is composed of individual cells separated by a crosswall, or **septum.** These hyphae are called **septate hyphae.** A few fungi, including *Rhizopus,* shown in Figure 35.1a, have hyphae that lack septa and are a continuous mass of cytoplasm with multiple nuclei. These are called **coenocytic hyphae.**

Until recently, fungi were classified on the basis of their sexual spores. Fungi that do not produce sexual spores were put in a "holding" phylum called **Deuteromycota.** Now fungi are grouped into phyla based on similarities in their rRNA. In a laboratory, fungi are identified by the appearance of their colony (color, size, and so on), hyphal organization (septate or coenocytic), and the structure and organization of reproductive

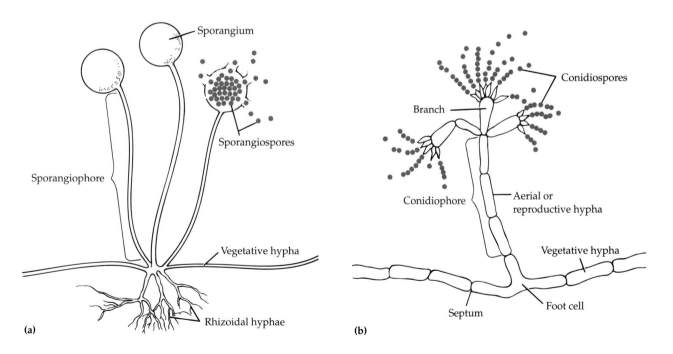

Figure 35.1

Asexual spores are produced by aerial hyphae. **(a)** Sporangiospores are formed within a sporangium. **(b)** Conidiospores are formed in chains. One possible arrangement is shown here.

spores. Because of the importance of colony appearance and organization, culture techniques and microscopic examination of fungi are very important.

Members of the phylum **Zygomycota** are saprophytic molds that have coenocytic hyphae. A **saprophyte** obtains its nutrients from dead organic matter; in healthy animals and plants, they do not usually cause disease. A common example is *Rhizopus* (bread mold). The asexual spores are formed inside a **sporangium,** or spore sac, and are called **sporangiospores** (Figure 35.1a). Sexual spores called **zygospores** are formed by the fusion of two cells.

The **Oomycota** are found in aquatic habitats and form sexual spores called **oospores,** formed by the fusion of two cells, and motile asexual spores called **zoospores.**

The **Ascomycota** include molds with septate hyphae and some yeasts. They are called sac fungi because their sexual spores, called **ascospores,** are produced in a sac, or **ascus.** The saprophytic molds usually produce **conidiospores** asexually (Figure 35.1b). The arrangements of the conidiospores are used to identify these fungi. Common examples are *Penicillium* and *Aspergillus*.

The **Basidiomycota** include the fleshy fungi, or mushrooms, and have sexual spores called **basidiospores** produced by the club-shaped cap called a **basidium**.

A classification scheme for saprophytic fungi is shown in Table 35.1. A majority of the fungi classified in Table 35.1 are saprophytes, but each division contains a few genera that are **pathogens,** causing disease in plants and animals.

Fungi, especially molds, are important clinically and industrially. Spores in the air are also the most common source of contamination in the laboratory. Sabouraud agar is a selective medium that is commonly used to isolate fungi.

Some pathogenic fungi exhibit **dimorphism**—that is, they have two growth forms. In pathogenic fungi, dimorphism is usually temperature dependent. The fungus is a yeast at 37°C and a mold at 25°C.

In this exercise, we will examine different molds and demonstrate dimorphism. The dimorphism of *Mucor* is not temperature dependent; it will be left for you to decide what physical condition affects the growth of *Mucor*.

Materials

Mold Culture
Petri plates containing Sabouraud agar (2)

Melted Sabouraud agar

Vaspar (one-half petroleum jelly and one-half paraffin)

Sterile Petri dish

Coverslips

Bent glass rod

Pasteur pipette

Dimorphic Gradient
Sabouraud agar, 5 ml, melted at 48°C

5 ml plastic microbeakers or small paper cups (2)

Razor blade (second period)

Tape

Cultures (as assigned)

Rhizopus stolonifer

Aspergillus niger

Penicillium notatum

Mucor rouxii

Table 35.1

Characteristics of Common Saprophytic Fungi

Phylum	Growth Characteristics	Asexual Reproduction	Sexual Reproduction
Oomycota ("water molds")	Coenocytic hyphae	Motile sporangiospores called zoospores	Oospores
Zygomycota ("conjugation fungi")	Coenocytic hyphae	Sporangiospores	Zygospores
Ascomycota ("sac fungi")	Septate hyphae; yeastlike	Conidiospores; budding	Ascospores
Basidiomycota ("club fungi")	Septate hyphae; includes fleshy fungi (mushrooms)	Fragmentation	Basidiospores

Demonstrations

Prepared slides of the following:
Zygospores
Ascospores

Techniques Required

Compound light microscopy, Exercise 1

Dissecting microscope, Appendix E

Sabouraud agar, Exercise 34

Procedure

First Period

1. Contaminate one Sabouraud agar plate in any manner you desire. Expose it to the air (outside, hall, lab, or wherever) for 15 to 30 minutes, or touch it. Incubate the plate, inverted, at room temperature for 5 to 7 days.

2. Inoculate the other Sabouraud agar plate with the mold culture assigned to you. Make one line in the center of the plate. Why don't you streak it? _____

   Incubate the plate, inverted, at room temperature for 5 to 7 days. (See color plates IV.2, IV.3, and V.7.)

3. Set up a slide culture of your assigned mold (Figure 35.2).
   a. Clean a slide and dry it.
   b. Place a drop of Sabouraud agar on the slide, flatten it to the size of a dime, and let it solidify.
   c. Using a sterile hot loop, carefully scrape off half of the circle of agar, leaving a smooth edge.
   d. Gently shake the mold culture to resuspend it, and inoculate the straight edge of the agar with a loopful of mold.
   e. Place a coverslip on the inoculated agar.
   f. Using a Pasteur pipette, add melted vaspar to seal three edges. Do not seal the inoculated edge.
   g. Put a piece of wet paper towel in the bottom of a Petri plate. Why? _____

   Place the slide culture on the glass rod on top of the towel and close the Petri plate. Incubate it at room temperature for 2 to 5 days. Do *not* invert the plate. What is the purpose of the glass rod? _____

4. Observe the prepared slides showing sexual spores. Carefully diagram each of the spore formations in the spaces provided in the Laboratory Report.

(a)   Place agar medium on the slide.

(b)   Cut a straight edge on one side of the solidified agar.

(c)   Inoculate the mold onto the straight edge.

(d)   Place a coverslip on the agar.

Vaspar

(e)   Seal the three uninoculated edges with vaspar, and place the slide on a glass rod in a Petri plate.

Figure 35.2
Preparation of a slide culture.

Dimorphic Gradient*

1. Flick the fungal culture tube to resuspend the culture. Inoculate the melted agar with 2 or 3 loopfuls

*Adapted from S. Bertnicki-Garcia. "The Dimorphic Gradient of *Mucor rouxii:* A Laboratory Exercise." *ASM News* 38(9):486–488, 1972. Permission granted by the American Society for Microbiology.

Figure 35.3
Tape the two beakers together as shown.

(a)

(b)

Figure 35.4
Observing fungal growth microscopically. **(a)** Cut the bottom beaker in half with a razor blade, and cut a vertical slice of agar. **(b)** Place the thin slice of agar on a slide, and cover it with a coverslip.

of the *Mucor* culture. Mix the tube by rolling it between your hands, and quickly pour the contents into an empty beaker before they harden. Why doesn't the beaker need to be sterile? _____

_____

2. Place a piece of wet paper towel in the remaining empty beaker, and invert it over the beaker containing the agar (Figure 35.3). Tape the edges.
3. Incubate the taped beakers at room temperature until growth occurs (5 to 7 days).

Second Period
1. Examine plate cultures of *each* mold (without a microscope), and describe their color and appearance. Then examine them with a dissecting microscope. Look at both the top and the underside.
2. Examine your contaminated Sabouraud plate and describe the results.
3. Examine slide cultures of each mold using a dissecting microscope. Record your observations.
4. To examine your dimorphic gradient, remove the tape and cut the beaker in half vertically with a razor blade (Figure 35.4a).

5. Cut a thin (approximately 1 mm) vertical slice of the inoculated agar with a razor blade. Carefully place the slice of agar on a slide and cover it with a coverslip (Figure 35.4b). Discard the razor blade by placing it in the sharps container.
6. Observe the slide under low and high power. Scan the agar from the bottom of the slice to the top. Many focal planes are present.
7. Discard your beaker in the biohazard container and your slide in the autoclave container.

exercise 35

Fungi: Molds

Name _____

Date _____

Lab Section _____

Purpose _____

_____

_____

Observations

Plate Cultures

	Organism			
	Rhizopus stolonifer	*Aspergillus niger*	*Penicillium notatum*	Unknown from Contaminated Plate
Colony appearance				
Macroscopic Hyphae color				
Spore color				
Underside color				
Microscopic Diagram				

Slide Cultures

Characteristics	Organism		
	Rhizopus stolonifer	*Aspergillus niger*	*Penicillium notatum*
Coenocytic?			
Type of spores			
Color of spores			
Diagram hyphal and spore arrangement			

Prepared Slides

Sexual spores

   Zygomycete   Ascomycete

Genus: _____   Genus: _____

                          _____ ×   _____ ×

Dimorphic Gradient

Fungus: _____

Top of agar (draw the structures seen):

Magnification: _____ ×   _____ ×   _____ ×

Middle of agar:

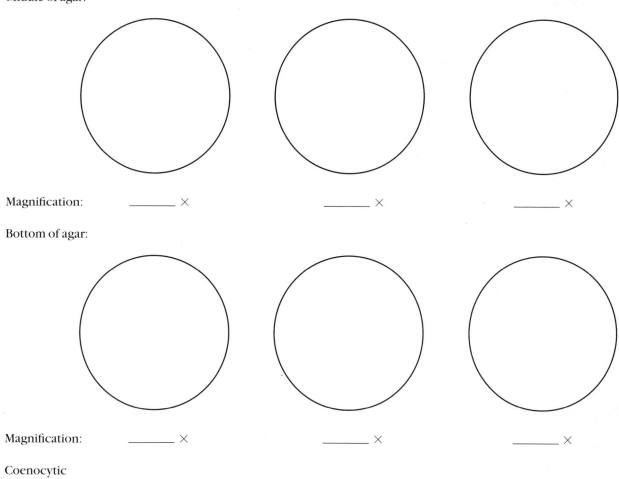

Magnification: _____ ✕   _____ ✕   _____ ✕

Bottom of agar:

Magnification: _____ ✕   _____ ✕   _____ ✕

Coenocytic
or septate? _____   _____   _____

Questions

1. Fill in the following table.

Organism	Asexual Spores	Sexual Spores	Phylum
Rhizopus			
Aspergillus			
Penicillium			

2. How do mold spores differ from bacterial endospores? _____

_____

_____

_____

3. What determines the form of *Mucor* seen in the dimorphic gradient? _____

    _____

    _____

    _____

4. Why do media used to culture fungi contain sugars? _____

    _____

5. Why are antibiotics frequently added to Sabouraud agar for isolation of fungi from clinical samples? _____

    _____

    _____

6. How can fungi cause respiratory-tract infections? _____

    _____

    _____

7. Why weren't pathogenic fungi used in this exercise? _____

    _____

    _____

Critical Thinking

An 82-year-old man was hospitalized with pneumonia. He was treated with penicillin and discharged after 7 days when his symptoms subsided. Two weeks later, the man was readmitted to the hospital for pneumonia. No bacteria or viruses were cultured from a biopsy specimen. Broad-spectrum antibiotic therapy was administered. The patient developed pneumonia again after returning home from an 8-day hospital stay. A bronchoscopy showed septate hyphae in his lung tissue. An investigation of his home showed that he did not keep pets but some sparrows had nested outside the living room window; he was not a gardener or outdoorsman. *Aspergillus* was cultured from his humidifier.

1. What is your preliminary identification of the pathogen?

2. Why was the diagnosis delayed?

3. Why did his infection recur?

exercise 36

Phototrophs: Algae and Cyanobacteria

Objectives

After completing this exercise, you should be able to:

1. List criteria used to classify algae.
2. List general requirements for the growth of phototrophic organisms.
3. Compare and contrast algae with fungi and bacteria.

Background

Most freshwater phototrophs belong to the groups listed in Table 36.1. Of primary interest to microbiologists are the **cyanobacteria.** Cyanobacteria have prokaryotic cells and belong to the Domain Bacteria.

Algae is the common name for photosynthetic eukaryotic organisms that lack true roots, stems, and leaves. Algae may be found in the ocean and in freshwater and on moist tree bark and soil. (See color plate IV.4.) Algae may be unicellular, colonial, filamentous, or multicellular. They exhibit a wide range of shapes: from the giant brown algae or kelp and delicate marine red algae to spherical green-algal colonies. Algae are classified according to pigments, storage products, the chemical composition of their cell walls, and flagella.

While the growth of phototrophs is essential in providing oxygen and food for other organisms, some filamentous algae, such as *Spirogyra,* are a nuisance to humans because they clog filters in water systems (see color plate IV.5). Phototrophs can be used to determine the quality of water. Polluted waters containing excessive nutrients from sewage or other sources have more cyanobacteria and fewer diatoms than clean waters do. Additionally, the *number* of algal cells indicates water quality. More than 1,000 algal cells per milliliter indicates that excessive nutrients are present.

Materials

Pond water samples:

A. Incubated in the light for 4 weeks.
B. Incubated in the dark for 4 weeks.
C. With nitrates and phosphates added; incubated in the light for 4 weeks.
D. With copper sulfate added; incubated in the light for 4 weeks.

Table 36.1

Some Characteristics of Major Groups of Phototrophs Found in Freshwater

Characteristics	Bacteria	Algal Protists		
	Cyanobacteria	Euglenoids	Diatoms	Green Algae
Color	Blue-green	Green	Yellow-brown	Green
Cell wall	Bacteria-like	Lacking	Readily visible with regular markings	Visible
Cell type	Prokaryote	Eukaryote	Eukaryote	Eukaryote
Flagella	Absent	Present	Absent	Present in some
Cell arrangement	Unicellular or filamentous	Unicellular	Unicellular or colonial	Unicellular, colonial, or filamentous
Nutrition	Autotrophic	Facultatively heterotrophic	Autotrophic	Autotrophic
Produce O_2	Yes	Yes	Yes	Yes

Techniques Required

Compound light microscopy, Exercise 1

Hanging drop, Exercise 2

Procedure

1. Prepare a hanging-drop slide from a sample of pond water A. Take your drop from the bottom of the container. Why? _____
    _____

2. Examine the slide using the low and high-dry objectives. Identify the algae present in the pond water. Refer to color plate XIII for identification. Draw those algae that you cannot identify. Record the relative amounts of each type of alga from $4+$ (most abundant) to $+$ (one representative seen).

3. Repeat the observation and data collection for the remaining pond water samples.

exercise 36

Phototrophs: Algae and Cyanobacteria

Name _____

Date _____

Lab Section _____

Purpose _____

_____

Data

Name of Alga or Cyanobacterium	Relative Abundance in Pond Water			
	Incubated in Light	Incubated in Dark	Incubated with NO_3^- and PO_4^{3-}	Incubated with $CuSO_4$
Drawings of other algae seen				
Total number of species				
Total number of organisms				

Which sample is the control? _____

Conclusions

Compare the number and various groups of algae observed in each environment with the control.

Dark vs. light (sample A)

In which sample did you expect more algae? _____

Why? _____

_____

_____

Did your findings agree with expected results? If not, briefly explain why. _____

_____

_____

Nitrate and phosphate vs. no nitrate and phosphate (sample A)

In which sample did you expect more algae? _____

Why? _____

_____

_____

Did your findings agree with expected results? If not, briefly explain why. _____

_____

_____

Copper vs. no copper (sample A)

In which sample did you expect more algae? _____

Why? _____

_____

_____

Did your findings agree with expected results? If not, briefly explain why. _____

_____

_____

Questions

1. Why can algae and cyanobacteria be considered indicators of productivity as well as of pollution? _____

    _____

2. How can algae be responsible for the production of more oxygen than land plants? _____

    _____

    _____

3. Why aren't algae included in the Kingdom Plantae? _____

    _____

    _____

    _____

4. Describe one way in which algae and fungi differ. _____

    _____

    _____

    How are they similar? _____

    _____

    _____

    _____

Critical Thinking

1. Outbreaks of cyanobacterial intoxication associated with lakes and ponds are reported annually. What would cause an increased number of cases in summer months? Why aren't cyanobacterial intoxications associated with swimming pools?

2. Cyanobacteria were once called "blue-green algae." What characteristics would lead to the name "blue-green algae"? What caused biologists to reclassify them as cyanobacteria?

exercise 37

Protozoa

Objectives

After completing this exercise, you should be able to:

1. List three characteristics of protozoa.
2. Explain how protozoa are classified.

Background

Protozoa are unicellular eukaryotic organisms that are loosely grouped as protists. Many protozoa live in soil and water, and some are normal microbiota in animals. A few species of protozoa are parasites.

Protozoa are heterotrophs, and most are aerobic. They feed on other microorganisms and on small particulate matter. Protozoa lack cell walls; in some, the outer covering is a thick, elastic membrane called a **pellicle.** Cells with a pellicle require specialized structures to take in food. The **contractile vacuole** may be visible in some specimens. This organelle fills with freshwater and then contracts to eliminate excess freshwater from the cell, allowing the organism to live in low-solute environments. Would you expect more contractile vacuoles in freshwater or marine protozoa? _____

In this exercise, we will examine free-living members of three phyla of protozoa. The **Rhizopoda** phylum consists of protozoa that move by using pseudopods. The **amoebas** (Figure 37.1a) move by extending lobelike projections of cytoplasm called **pseudopods.** As pseudopods flow from one end of the cell, the rest of the cell flows toward the pseudopods.

The **Euglenozoa** (Figure 37.1b) have one or more flagella. Although flagellates are heterotrophs, the organism used in this exercise is a facultative heterotroph. It grows photosynthetically in the presence of light and heterotrophically in the dark.

Members of the phylum **Ciliophora** (Figure 37.1c) have many cilia extending from the cell. In some ciliates, the cilia occur in rows over the entire surface of the cell. In ciliates that live attached to solid surfaces, the cilia occur only around the oral groove. Why only around the oral groove? _____

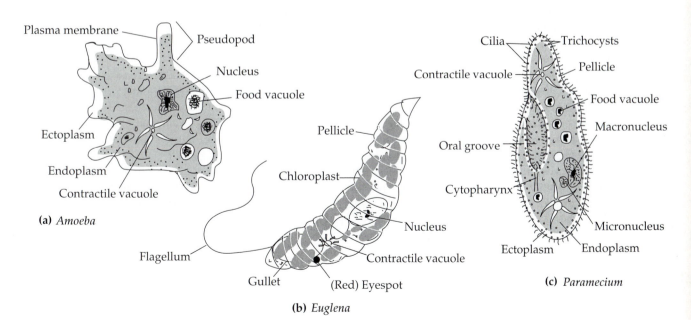

(a) *Amoeba*

(b) *Euglena*

(c) *Paramecium*

Figure 37.1

Protozoa. **(a)** *Amoeba* moves by extending pseudopods. **(b)** *Euglena* has a whiplike flagellum. **(c)** *Paramecium* has cilia over its surface.

Food is taken into the **oral groove** through the cytostome (mouth) and into the cytopharynx, where a **food vacuole** forms.

Materials

Methylcellulose, 1.5%

Acetic acid, 5%

Pasteur pipettes

Toothpicks

Coverslips

Cultures

Amoeba

Paramecium

Paramecium feeding on Congo red–yeast suspension

Euglena

Techniques Required

Compound light microscopy, Exercise 1

Hanging drop, Exercise 2

Procedure

1. Prepare a wet mount of *Amoeba*. Place a drop from the bottom of the *Amoeba* culture on a slide. Place one edge of the coverslip into the drop, and let the fluid run along the coverslip (Figure 37.2). Gently lay the coverslip over the drop. Observe the amoeboid movement, and diagram it. Which region of the cytoplasm has more granules: the ectoplasm or the endoplasm? (Refer to Figure 37.1a.)

2. Prepare a wet mount of *Euglena*. Follow one individual and diagram its movement. Can you see *Euglena*'s red "eyespot"? _____
    Why do you suppose it is present in photosynthetic strains and not in nonphotosynthetic strains? _____

    _____

    Allow a drop of acetic acid to seep under the coverslip (Figure 37.3). How does *Euglena* respond?

    _____

Figure 37.2
Gently lower the coverslip.

Figure 37.3
Add a drop of acetic acid at one edge of the coverslip. Allow it to diffuse into the wet mount.

3. Prepare a wet mount of *Paramecium* and observe its movement. (See color plate IV.5.) Why do you suppose it rolls and *Amoeba* does not? _____

4. Place a drop of methylcellulose on a slide. Make a wet mount of the *Paramecium* culture that has been feeding on a Congo red–yeast suspension in this mixture. The *Paramecium* will move more slowly in the viscous methylcellulose. Observe the ingestion of the red-stained yeast cells by *Paramecium*. Congo red is a pH indicator. As the contents of food vacuoles are digested, the indicator will turn blue (pH 3). What metabolic products would produce acidic conditions in the vacuoles? _____

    _____

    Count the number of red and blue food vacuoles in a *Paramecium*. Sketch a *Paramecium* and identify the locations of the food vacuoles.

exercise 37

Protozoa

Name _____

Date _____

Lab Section _____

Purpose _____

_____

_____

Observations

Use a series of diagrams to illustrate the following:

The movement of *Amoeba* across the field of vision. _____ ×

The movement of *Euglena* and its flagellum. _____ ×

The ingestion of food, formation of a food vacuole, and movement of food vacuoles in *Paramecium*. Note any color changes in the food vacuoles. _____ ✕

Questions

1. How did *Euglena* respond to the acetic acid? _____

_____

2. Which organism observed in this exercise would you bet on in a race? _____

3. Describe the arrangement of cilia on *Paramecium*. _____

_____

4. Match the characteristics (a–g) to the phyla listed. Name one genus from each phylum.

Phylum	Characteristics	Example Genus
Archaezoa		
Apicomplexa		
Dinoflagellates		
Ciliophora		
Euglenozoa		

Characteristics:
a. Apical enzymes
b. Cilia
c. Flagella
d. Have cellulose
e. No mitochondria
f. Photosynthetic
g. Pseudopods

Critical Thinking

1. Why is *Euglena* often used to study algae and protozoa?

2. *Pneumocystis* has been classified as a protozoan since 1908. However, rRNA sequencing suggests that it is a fungus. Why might it be important to have an accurate classification of this organism?

3. Laboratory eyewashes should be flushed once a month to remove *Acanthamoeba* accumulations from the pipes. Why is this removal necessary for eyewashes but not other water outlets?

part nine

Viruses

EXERCISES

Viruses are fundamentally different from the bacteria studied in previous exercises, as well as from all other organisms. Wendell M. Stanley, who was awarded the Nobel Prize in 1946 for his work on tobacco mosaic virus, described the virus as "one of the great riddles of biology. We do not know whether it is alive or dead, because it seems to occupy a place midway between the inert chemical molecule and the living organism."*

Viruses are submicroscopic, filterable, infectious agents (see the illustration on the next page). Viruses are too small to be seen with a light microscope and can be seen only with an electron microscope. Filtration is frequently used to separate viruses from other microorganisms. A suspension containing viruses and bacteria is filtered through a membrane filter with a small pore size (0.45 μm) that retains the bacteria but allows viruses to pass through.

Viruses are *obligate intracellular parasites* with a simple structure and organization. Viruses contain DNA or RNA enclosed in a protein coat. Viruses exhibit quite strict specificity for host cells. Once inside the host cell, the virus may use the host cell's synthetic machinery and material to cause synthesis of virus particles that can infect new cells.

Viruses are widely distributed in nature and have been isolated from virtually every eukaryotic and prokaryotic organism. Based on their host specificity, viruses are divided into general categories, such as **bacterial viruses (bacteriophages), plant viruses,** and **animal viruses.** A virus's host cells must be grown in order to culture viruses in a laboratory.

*Quoted in H. A. Lechevalier and M. Solotorovsky. *Three Centuries of Microbiology.* New York: Dover Publications, 1974, p. 25.

In Exercise 38, we will isolate bacteriophages and estimate the number of phage particles. In Exercise 39, we will isolate plant viruses in plant leaves.

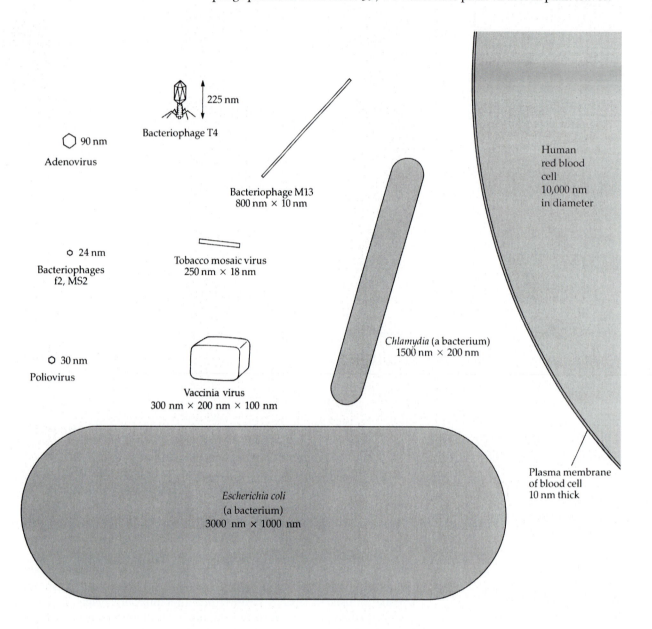

Comparative sizes of several viruses, bacteria, and a human red blood cell.

exercise 38

Isolation and Titration of Bacteriophages

Objectives

After completing this exercise, you should be able to:

1. Isolate a bacteriophage from a natural environment.
2. Describe the cultivation of bacteriophages.
3. Determine the titer of a bacteriophage sample using the broth-clearing and plaque-forming methods.

Background

Bacteriophages, a term coined around 1917 by Félix d'Hérelle meaning "bacteria eater," parasitize most, if not all, bacteria in a very specific manner. Some bacteriophages, or **phages,** such as T-even bacteriophages, have a **complex structure** (Figure 38.1a). The protein coat consists of a polyhedral head and a helical tail, to which other structures are attached. The head contains the nucleic acid. To initiate an infection, the bacteriophage **adsorbs** onto the surface of a bacterial cell by means of its tail fibers and base plate. The bacteriophage injects its nucleic acid into the bacterium during **penetration** (Figure 38.1b). The tail sheath contracts, driving the tail core through the cell wall and injecting the nucleic acid into the bacterium.

Bacteriophages can be grown in liquid or solid cultures of bacteria. The use of solid media makes location of the bacteriophage possible by the **plaque-forming method.** Host bacteria and bacteriophages are mixed together in melted agar, which is then poured into a Petri plate containing hardened nutrient agar. Each bacteriophage that infects a bacterium multiplies, releasing several hundred new viruses. The new viruses infect other bacteria, and more new viruses are produced. All the bacteria in the area surrounding the original virus are destroyed, leaving a clear area, or **plaque,** against a confluent "lawn" of bacteria. The lawn of bacteria is produced by the growth of uninfected bacterial cells.

In this exercise, we will isolate a bacteriophage from host cells in a natural environment (i.e., in sewage or houseflies). Since the numbers of phages in a natural source are low, the desired host bacteria and additional nutrients are added as an **enrichment procedure.** After incubation, the bacteriophage can be isolated by centrifugation of the enrichment media and membrane filtration. **Filtration** has been used for removing mi-

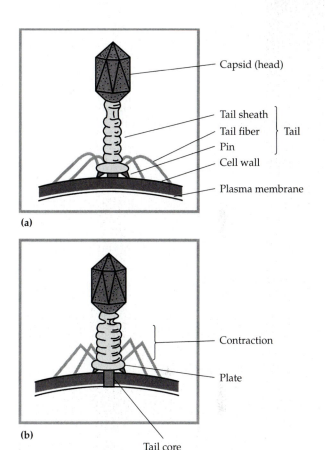

Figure 38.1

T-even bacteriophage. **(a)** Diagram of a bacteriophage showing its component parts and adsorption onto a host cell. **(b)** Penetration of a host cell by the bacteriophage.

crobes from liquids for purposes of sterilization since 1884, when Pasteur's associate, Charles Chamberland, designed the first filter candle. Today most filtration of viruses is done using membrane filters with pore sizes (usually 0.45 μm) that physically exclude bacteria from the filtrate.

You will measure the viral activity in your sample by performing sequential dilutions of the viral preparation and assaying for the presence of viruses. In the **broth-clearing assay,** the **endpoint** is the highest dilution (smallest amount of virus) producing lysis of bacteria and clearing of the broth. The **titer,** or concentration, that results in a recognizable effect is the reciprocal of the endpoint. In the plaque-forming method, the titer is determined by counting plaques. Each plaque theoretically corresponds to a single infective virus in the initial suspension. Some plaques may arise from more than one virus particle, and some virus particles may not be infectious. Therefore, the titer is determined by counting the number of **plaque-forming units (pfu).** The titer, plaque-forming units per milliliter, is determined by counting the number of plaques and dividing by the amount plated times the dilution. For example, 32 plaques with 0.1 ml plated of a $1:10^3$ dilution is equal to

$$\frac{32}{0.1 \times 10^{-3}} = 3.2 \times 10^5 \text{ pfu/ml}$$

In this exercise, we will determine the viral activity by a plaque-forming assay and a broth-clearing assay.

Materials

Isolation of Bacteriophage
Use flies, sewage, or bacteriophage culture.

   Flies as source:
   Houseflies (fresh) (20 to 25); bring your own (Figure 38.2)
   Trypticase soy broth (20 ml)
   Mortar and pestle

   Sewage as source:
   Raw sewage (45 ml)
   10× nutrient broth (5 ml)
   50 ml graduated cylinder
   Funnel

   Bacteriophage as source:
   Bacteriophage
   Trypticase soy broth (20 ml)

Sterile 125 ml Erlenmeyer flask

5 ml sterile pipette (1)

Centrifuge tubes

Screw-capped tube

Sterile membrane filter (0.45 μm)

Sterile membrane filter apparatus

Safety goggles

Gloves

Titration of Bacteriophage
Tubes containing 9 ml trypticase soy broth (6)

Petri plates containing nutrient agar (6)

Tubes containing 3 ml melted soft trypticase soy agar (0.7% agar) (6)

1 ml sterile pipettes (7)

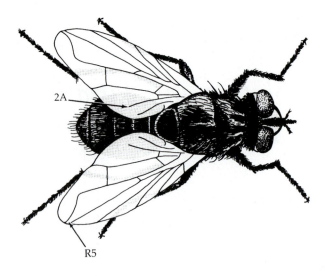

Figure 38.2
Houseflies may harbor bacteriophages. To identify a housefly, observe that the 2A vein doesn't reach the wing margin, and the R5 cell narrows.

Cultures

Escherichia coli broth

Bacteriophage culture or bacteriophage from isolation procedure.

Techniques Required

Pour plate, Exercise 11

Pipetting, Appendix A

Serial dilution technique, Appendix B

Membrane filtration, Appendix F

Procedure

> **Wear gloves and goggles when working with raw sewage. Handle flies carefully.**

Isolation of Bacteriophage
1. Follow enrichment procedure a, b, or c.
    a. Wear gloves. Add the flies to half of the trypticase soy broth in the mortar, and grind with the pestle to a fine pulp. Transfer this mixture to a sterile flask, and add the remainder of the broth and 2 ml *E. coli*. Place the mortar and pestle in the To Be Autoclaved basket, and the gloves in the biohazard container.
    b. Wear gloves and safety glasses. Add 45 ml sewage to 5 ml 10× nutrient broth in a sterile flask. Add 5 ml *E. coli* broth. Mix gently.

Place gloves in the biohazard container. Why is the broth ten times more concentrated than normal? _____

c. Add 0.5 ml of bacteriophage to 20 ml of trypticase soy broth in a sterile flask. Add 2 ml *E. coli* broth. Mix gently.

2. Incubate the enrichment for 24 hours at 35°C.

3. Decant 10 ml of the enrichment into a centrifuge tube. Place the tube in a centrifuge.

> ⚠ **Balance the centrifuge with a similar tube containing 10 ml water.**

Centrifuge at 2500 rpm for 10 minutes to remove most bacteria and solid materials.

4. Filter the supernatant through a membrane filter. Decant the clear liquid into a screw-capped tube. How does filtration separate viruses from bacteria? _____

Store at 5°C until the next lab period.

Titration of Bacteriophage (Figure 38.3)
Read carefully before proceeding.

1. Label the plates and the broth tubes "1" through "6."

2. Aseptically add 1 ml phage suspension (from step 4, above) to tube 1. Mix by carefully aspirating up and down three times with the pipette. Using a different pipette, transfer 1 ml to the second tube, mix well, and then put 1 ml into the third tube. Why is the pipette changed? _____
Continue until the fifth tube. After mixing this tube, discard 1 ml into a container of disinfectant. What dilution exists in each tube? (See Figure 38.3.)

	Tube 1	Tube 2	Tube 3
Dilutions:	____	____	____

	Tube 4	Tube 5	Tube 6
Dilutions:	____	____	____

What is the purpose of tube 6? _____

3. With a pipette, add 0.1 ml *E. coli* to the soft agar tubes and place them back in the water bath. Keep the water bath at 43°C to 45°C at all times. Why? _____

4. With the remaining pipette, start with broth tube 6 and aseptically transfer 0.1 ml from tube 6 to the soft agar tube. Mix by swirling, and quickly pour the inoculated soft agar evenly over the surface of Petri plate 6. Then, using the same pipette, transfer 0.1 ml from tube 5 to a soft agar tube, mix, and pour over plate 5. Continue until you have completed tube 1. Why can the procedure be done with one pipette? _____

5. After completing step 4, add 2 loopfuls of *E. coli* to each of the remaining broth tubes and mix. Incubate the tubes at 35°C. Observe them in a few hours. In the broth tubes, record the highest dilution that was clear as the endpoint. The titer is the reciprocal of the endpoint.

6. Incubate the plates at 35°C until plaques develop. The plaques may be visible within hours. (See color plate VII.1.) Plates can be stored at 5°C.

7. Select a plate with between 25 and 250 plaques. Count the number of plaques, and calculate the number of plaque-forming units (pfu) per milliliter. Record your results.

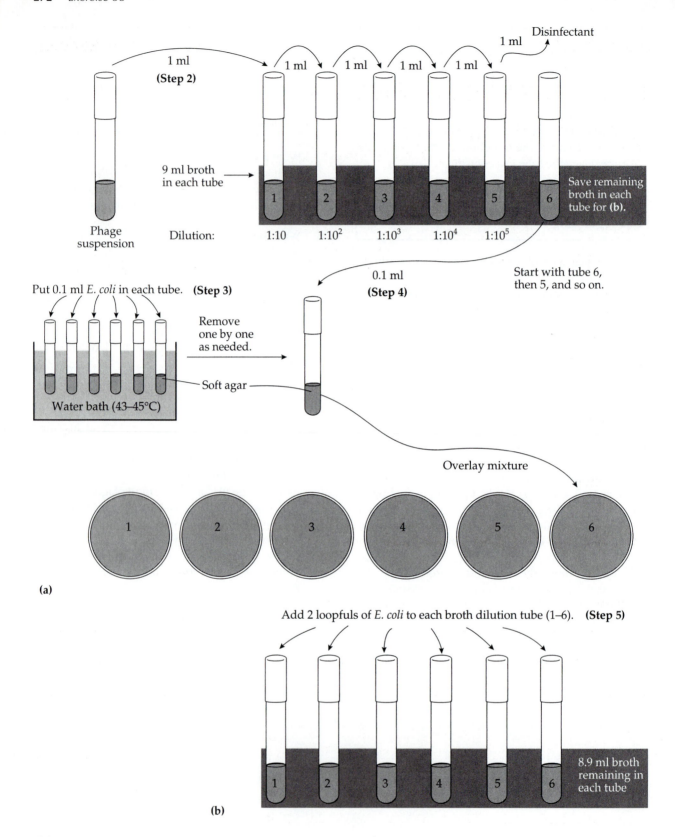

Figure 38.3
Procedure for titration of bacteriophages. **(a)** Plaque-forming assay. **(b)** Broth-clearing assay.

Isolation and Titration of Bacteriophages

Name _____

Date _____

Lab Section _____

Purpose _____

_____

_____

Data

Plaque-Forming Assay

Choose one plate with 25 to 250 plaques.

Draw what you observed.

   Number of plaques = _____

   Dilution used for that plate = _____

Broth-Clearing Assay

Indicate whether each tube was turbid or clear.

Incubated _____ hours.

Tube	Turbid or Clear	Dilution
1		
2		
3		
4		
5		
6		

Conclusions

Plaque-Forming Assay

pfu/ml = _____
Show your calculations.

Broth-Clearing Assay

What was the endpoint? _____

What was the titer? _____

Questions

1. Are any contaminating bacteria present? How can you tell? _____
_____

2. Why did you add *Escherichia coli* to sewage, which is full of bacteria? _____
_____
_____

3. Are all the plaques the same size? Briefly explain why or why not. _____
_____
_____

4. What does the endpoint represent in a broth-clearing assay? _____
_____
_____
_____

5. Which of these assays is more accurate? Briefly explain. _____
_____
_____

Critical Thinking

1. How would you develop a pure culture of a phage?

2. If there were no plaques on your plates, offer an explanation. How would you explain turbidity in all of the tubes in a broth-clearing assay?

3.   How would you isolate a bacteriophage for a species of *Bacillus*?

4.   Titration of a phage on a pure bacterial lawn resulted in some cloudy plaques. Briefly explain this result.

exercise 39

Plant Viruses

*Smoking was definitely dangerous to your
health in seventeenth-century Russia. Czar
Michael Federovitch executed anyone on
whom tobacco was found. But Czar Alexei
Mikhailovitch was easier on smokers; he
merely tortured them until they told
who their suppliers were.*

ANONYMOUS

Objectives

After completing this exercise, you should be able to:

1. Isolate a plant virus.
2. Describe the cultivation of a plant virus.
3. Determine the host range of a plant virus.

Background

Plant viruses are important economically in that they are the second leading cause of plant disease. (Fungi are the leading cause of plant disease.) In order for most plant viruses to infect a plant, the virus must enter through an abrasion in the leaf or stem. Insects called **vectors** carry viruses from one plant to another. Most viruses cause either a **localized infection,** in which the leaf will have necrotic lesions (brown plaques), or a **systemic infection,** in which the infection will run throughout the entire plant. One of the most studied viruses is tobacco mosaic virus (TMV), which was the first virus to be purified and crystallized.

In this experiment, we will attempt to isolate TMV from various tobacco products and determine its host range.

Materials

Tobacco plant

Tomato plant

Bean plant (*Chenopodium*)

Tobacco products of various types (bring some, if possible)

Mortar and pestle

Fine sand or Carborundum

Wash bottle of water

Paper labels or labeling tape

Cotton

Sterile water

Techniques Required

None.

Procedure

1. Select a plant, and label the pot with your name and lab section.
2. Place labels around the petioles (see Figure 39.1) of several leaves, with names corresponding to the various tobacco products available. Label one petiole "C" for control.
3. *Wash your hands carefully before and after each inoculation.*
4. Spray the control leaf with a small amount of water, and dust the leaf with Carborundum or sand.

> ⚠ **Be careful not to inhale Carborundum dust. It is a lung irritant.**

Gently rub the leaf with a cotton ball dampened in sterile water. Wash the leaf with a wash bottle of water. What is the purpose of this leaf? _____

_____

5. Select a tobacco product, place a "pinch" of it into a mortar, and add sterile water. Grind the mixture into a slurry with a pestle. Spray one leaf with water, dust it with Carborundum or sand, and gently rub it with a cotton ball soaked with the tobacco slurry. Wash the leaf with water. Record the name of the tobacco source.
6. Using the same tobacco product, repeat steps 1 through 5 on a different species of plant. Repeat steps 5 and 6 with another tobacco product, cleaning the mortar and pestle with soap and water between products.

7. Replace the plant in the rack.
8. Check the plant at each laboratory period for up to 3 to 4 weeks for brown plaques (localized infection) or wilting (systemic infection). Water your plants as needed. Discard plants in the To Be Autoclaved area. (See color plate VII.2.)

Figure 39.1
Label a plant. Place labels close to the leaves that will be treated, as shown.

exercise 39

Plant Viruses

Name _____

Date _____

Lab Section _____

Purpose _____

_____

_____

Data

Obtain data from your classmates for the other plant species and tobacco products.

Identify Plant Species and Tobacco Product Used		Appearance of Leaves							
		Lab Period							
		1	2	3	4	5	6	7	8
1.	Control								
	Test								
2.	Control								
	Test								
3.	Control								
	Test								
4.	Control								
	Test								
5.	Control								
	Test								
6.	Control								
	Test								

Draw a plant showing locations of the labeled leaves and any signs of infection.

Conclusions

What can you conclude about the host range of TMV? _____

_____

_____

_____

Questions

1. Did systemic disease occur? How can you tell? _____

_____

2. How do insects act as vectors of plant viruses? _____

_____

_____

3. If the control leaf was damaged, what happened? _____

_____

4. Why is the leaf rubbed with sand or Carborundum? _____

_____

Critical Thinking

1. Design an experiment to prove that damage to the test leaves was due to a viral infection.

2.   A reservoir of infection is a continual source of that infection. What is a likely reservoir for TMV?

3.   Describe how the titration of a plant virus would be accomplished.

4.   Why is smoking prohibited in greenhouses?

part ten

Interaction of Microbe and Host

EXERCISES

40   Epidemiology
41   Koch's Postulates

In 1883, after detecting microorganisms in the environment, Robert Koch asked:

> *What significance do these findings have? Can it be stated that air, water, and soil contain a certain number of microorganisms and yet are without significance to health?**

Koch felt that the causes, or **etiologic agents,** of infectious diseases could be found in the study of microorganisms. He proved that anthrax and tuberculosis were caused by specific bacteria, and his work provided the framework for the study of the etiology of any infectious disease.

Today we refer to Koch's protocols for identifying etiologic agents as **Koch's postulates** (see the illustration on the next page; also see Exercise 41). The study of how and when diseases occur is called **epidemiology** (Exercise 40).

*Quoted in R. N. Coetsch, ed. *Microbiology: Historical Contributions from 1776 to 1908.* New Brunswick, NJ: Rutgers University Press, 1960, p. 130.

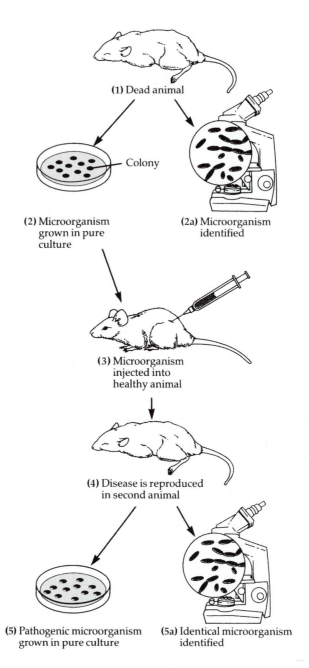

(1) Dead animal

Colony

(2) Microorganism
grown in pure
culture

(2a) Microorganism
identified

(3) Microorganism
injected into
healthy animal

(4) Disease is reproduced
in second animal

(5) Pathogenic microorganism
grown in pure culture

(5a) Identical microorganism
identified

Koch's postulates. (1) Microorganisms from a diseased
or dead animal are **(2)** grown in pure culture and
(2a) identified, if possible. **(3)** A sample of the pure culture
is inoculated into a healthy animal, and **(4)** the disease is
reproduced in the animal. **(5)** Pathogenic microorganisms
are again grown in pure culture and **(5a)** identified,
confirming that they are identical with the original
pathogen.

exercise 40

Epidemiology

Objectives

After completing this exercise, you should be able to:

1. Define the following terms: epidemiology, epidemic, reservoir, and carrier.
2. Describe three methods of transmission.
3. Determine the source of a simulated epidemic.

Background

In every infectious disease, the disease-producing microorganism, the **pathogen,** must come in contact with the **host,** the organism that harbors the pathogen. **Communicable diseases** can be spread either directly or indirectly from one host to another. Some microorganisms cause disease only if the body is weakened or if a predisposing event such as a wound allows them to enter the body. Such diseases are called **noncommunicable diseases**—that is, they cannot be transmitted from one host to another. The science that deals with when and where diseases occur and how they are transmitted in the human population is called **epidemiology. Endemic diseases** such as pneumonia are constantly present in the population. When many people in a given area acquire the disease in a relatively short period of time, it is referred to as an **epidemic disease.** The first reported patient in a disease outbreak is the **index case.**

Diseases can be transmitted by **direct contact** between hosts. **Droplet infection,** when microorganisms are carried on liquid drops from a cough or sneeze, is a method of direct contact. Diseases can also be transmitted by contact with contaminated inanimate objects, or **fomites.** Drinking glasses, bedding, and towels are examples of fomites that can be contaminated with pathogens from feces, sputum, or pus.

Some diseases are transmitted from one host to another by vectors. **Vectors** are insects and other arthropods that carry pathogens. In **mechanical transmission,** insects carry a pathogen on their feet and may transfer the pathogen to a person's food. For example, houseflies may transmit typhoid fever from the feces of an infected person to food. Transmission of a disease by an arthropod's bite is called **biological transmission.** An arthropod ingests a pathogen while biting an infected host. The pathogen can multiply or mature in the arthropod and then be transferred to a healthy person in the arthropod's feces or saliva.

The continual source of an infection is called the **reservoir.** Humans who harbor pathogens but who do not exhibit any signs of disease are called **carriers.**

An **epidemiologist** compiles data on the incidence of a disease and its method of transmission and tries to locate the source of infection in order to decrease the incidence.

In this exercise, an epidemic will be simulated. You will be the epidemiologist who, by deductive reasoning and with luck, determines the source of the epidemic.

Materials

Petri plate containing nutrient agar

Small plastic sandwich bag

1 unknown swab per student (the swab has *Serratia marcescens, Rhodotorula rubra,* or *Micrococcus luteus* on it. Your instructor will tell you which organism.)

Techniques Required

Colony morphology, Exercise 9

Procedure

1. Divide the Petri plate into five sectors, labeled "1" through "5."

Carefully read steps 2 through 4 before proceeding.

2. Put the plastic bag on your left hand. Carefully unwrap your swab without touching the cotton. Holding the swab with your right hand, rub it on the palm of your left hand (i.e., on the plastic bag). Discard the swab in a container of disinfectant.
3. Shake hands (using your left hand) with a classmate when the instructor gives the signal. Shake hands so your fingers touch the other's palm and vice versa. After shaking hands, touch your fingers to the first sector of the nutrient agar. Record the person's name and swab number.
4. Repeat step 3, shaking hands with four other classmates. Remember to touch your fingers to the corresponding sector of the nutrient agar after each handshake. Keep good records.

5. Discard the plastic bag in the To Be Autoclaved basket. Incubate the plate, inverted, at room temperature.

6. At the next lab period, deductively try to determine who had the contaminated fomite. (See color plates V.1 and XII.1.)

exercise 40

Epidemiology

Name _____

Date _____

Lab Section _____

Purpose _____

_____

_____

Data

Which organism was on the contaminated swab? _____

Results

Sector	Student's Name	Swab Number	Appearance of Colonies on Nutrient Agar
1.			
2.			
3.			
4.			
5.			

Attach your group's or the class's data.

Conclusions

1. Who had the contaminated swab? _____

2. What number was the contaminated swab? _____ Explain how you arrived at your conclu-

   sion. _____

   _____

3. Diagram the path of the epidemic.

Questions

1. Could you be the "infected" individual and not have growth on your plate? Explain. _____

_____

_____

_____

2. Do all people who contact an infected individual acquire the disease? _____

_____

_____

_____

3. How can an epidemic stop without chemotherapy and vaccines? _____

_____

_____

_____

4. Are any organisms other than the culture assigned for this experiment growing on the plates? How can you tell?

_____

_____

_____

5. What was the method of transmission of the "disease" in this experiment? _____

_____

_____

Critical Thinking

1. During a 3-month period, acute hepatitis B virus (HBV) infection was diagnosed in nine residents of a nursing home. Serological testing of all residents revealed that nine people had acute HBV infection, two had chronic infection, five were immune, and 58 were susceptible. Medical charts of residents were reviewed for history of medications and use of ancillary medical services. Infection-control practices at the nursing home were assessed through interviews with personnel and direct observations of nursing procedures. A summary of the medical charts is shown below.

	Case Patients	Susceptible Residents
Received insulin injections	11	58
Patients having capillary blood taken by fingersticks	11	6
≥60 fingersticks/month	7	0
<60 fingersticks/month	4	6
Average number of venous blood draws/month	23	6
Patients having both capillary and venous blood drawn	11	39
Visited by dentist	1	5
Visited by podiatrist	10	52
Blood transfusion	0	0

What is the usual method of transmission for hepatitis B? _____

What is the probable source of infection in hospitals? _____

How was *this* infection transmitted? _____

2. During a 2-week period, 13 of 56 patients at one hospital developed postoperative *Staphylococcus aureus* bacteremia following different surgical procedures. Epidemiological investigation identified that the intravenous anesthetic propofol was used, and that five people were probably present at all the surgeries. The propofol was delivered via a 6 ml syringe in a pump, and propofol remaining at the end of one surgery was used for the next surgical procedure. The results of laboratory cultures are shown below.

	S. aureus Phage Group II	
	Before Use	After Use
Propofol	0 bacteria/ml	10^3 bacteria/ml
Operating Room Staff	Face	Hands
Surgical technician	*S. aureus* phage group 1	—
Surgeon	—	*S. aureus* phage group I
Nurse anesthetist	—	*S. aureus* phage group II
Surgical nurse	—	—

What is the method of transmission of these infections? _____

How could these infections have been prevented?

3. A health department received a report from hospital A that 15 patients had been admitted on October 12 with unexplained pneumonia. On October 21, hospital B, located 15 miles from hospital A, reported a higher than normal pneumonia census for the first two weeks of October. *Legionella pneumophila* was eventually identified in 23 patients, 22 were hospitalized, and two died. To identify potential exposures associated with *L. pneumophila,* a questionnaire was developed, and a case-control study was initiated on November 2 to identify the source of infection. Three controls were selected for each confirmed case; controls were matched by age, gender, and underlying medical conditions. Of the 15 case patients for whom a history was available, 14 had visited a large home improvement center 2 weeks before onset of illness. Results of the questionnaire are shown below.

	Case Patients	Control
Number of patients	15	45
Visited home improvement center	14	12
Average time at center (min)	79	29
Looked at whirlpool spa X	13	9
Looked at whirlpool spa Y	13	1
Visited greenhouse sprinkler system display	10	10
Visited decorative fish pond	14	12
Used drinking fountain	13	10
Used urinals	10	4
Used restroom hot water	6	4
Used restroom cold water	8	8

What is the most likely source of this outbreak of legionellosis? _____

How would you prove this was the source?

How was this disease transmitted?

Provide an explanation of the infected patient who did not go to the home improvement center.

4. Thirty-two people attended a brunch (11 A.M. to 3 P.M.) on May 22, in Livonia, New York. All fresh food items were delivered to the cooking area the morning of May 22. The menu consisted of omelets, bacon, fruit salad, and pastries.

Omelets consisted of grade A eggs from a Maryland farm. They were beaten and mixed with milk, salt, pepper, and diced onions. The egg batter was fried in small batches and served from warming trays heated by alcohol burners.

Bacon was purchased from a meat wholesaler. The bacon was fried on a large griddle and served from warming trays.

Fruit salad was made from fresh apples, bananas, and oranges; they were cut up and mixed with commercially canned peaches and pears.

Muffins and other breads were purchased from a bakery and served with butter.

Case	Foods Eaten[a] 1	2	3	4	Beverages[b]	Time of Meal	Symptoms[c]	Onset of Symptoms Day	Hr
1	x	x	x	x	C, J	1100	D, V, N, A	Sun	2400
2		x	x	x	M, J	1200			
3	x	x	x	x	C	1200	N, A	Mon	0200
4	x	x	x	x	C	1200	D, N, A	Mon	0600
5				x	T	1100			
6		x		x	C, J	1200			
7			x	x	C, J	1300			
8				x	C, J	1300			
9	x	x	x	x	C	1100	D, V, N, A	Mon	1100
10				x	T	1200			
11				x	M	1300			
12	x	x	x	x	M	1300	D, N, A	Mon	1400
13				x	C	1200			
14		x	x	x	C, J	1500			
15	x	x		x	T	1500	D, N, V	Mon	2400
16				x	M, J	1100			
17		x	x	x	C	1100			
18		x		x	C	1200	N	Sun	1300
19	x	x			T	1200	D, V, N, A	Mon	1700
20			x	x	C, J	1200			
21	x	x		x	M	1300	D, N, A	Mon	0200
22				x	M	1400	A	Tue	0800
23				x	M, J	1400			
24	x	x	x	x	C	1100	D, A	Mon	1500
25	x	x		x	C	1100	V, N, A	Mon	1100
26			x	x	T	1200			
27				x	T	1100			
28	x	x	x	x	C	1200	D, A	Mon	0200
29		x		x	M	1300			
30					C, J	1400			
31		x	x		C	1100			
32	x	x	x	x		1200	D, A	Mon	1200

[a]Foods Eaten: 1, omelet; 2, bacon; 3, fruit salad; 4, pastry.
[b]Beverages: C, coffee; T, tea; M, milk; J, orange juice.
[c]Symptoms: D, diarrhea; N, nausea; V, vomiting; A, abdominal cramps.

What was the source of this disease? _____

How was it transmitted? _____

Briefly explain how you arrived at your conclusion.

exercise 41

Koch's Postulates

Objectives

After completing this exercise, you should be able to:

1. Define the following terms: etiologic agent, pathogenicity, and virulence.
2. List and explain Koch's postulates.

Background

The **etiologic agent** is the cause of an infectious disease. Microorganisms are the etiologic agents of a wide variety of infectious diseases in all forms of life. Microbes that cause diseases are called pathogens, and the process of disease initiation and progress is called **pathogenesis.** The interaction between the microbe and host is complex. Whether or not a disease occurs depends on our vulnerability or **susceptibility** to the pathogen and on the virulence of the pathogen. **Virulence** is the degree of pathogenicity. Factors influencing virulence include the number of pathogens, the portal of entry into the host, and toxin production.

The actual cause of many diseases is hard to determine. Although many microorganisms can be isolated from a diseased tissue, their presence does not prove that any or all of them caused the disease. A microbe may be a secondary invader or part of the normal microbiota or transient microbiota of that area. While working with anthrax and tuberculosis, Robert Koch established four criteria, now called **Koch's postulates,** to help identify a particular organism as the causative agent for a particular disease. Koch's postulates are the following:

1. The same organism must be present in every case of the disease.
2. The organism must be isolated from the diseased tissue and grown in pure culture in the laboratory.
3. The organism from the pure culture must cause the disease when inoculated into healthy, susceptible laboratory animals.
4. The organism must again be isolated—this time from the inoculated animals—and must be shown to be the same pathogen as the original organism.

These criteria are used by most investigators, but they cannot be applied to all infectious diseases. For example, viruses cannot be cultured on artificial media, and they are not readily observable in a host. Moreover, many viruses cause diseases only in humans and not in laboratory animals.

In this exercise, we will demonstrate Koch's postulates with one of two different bacteria: *Bacillus thu-*

ringiensis isolated from Milky Spore Disease of tomato hornworm larvae, or *Erwinia carotovora* isolated from soft rot of carrots.

Materials
Bacillus: Milky Spore Disease
Petri plate containing nutrient agar

Petri plate containing starch agar

Second period:

Scalpel and scissors

Tube containing sterile nutrient broth

Tomato hornworm larvae (2)

Sterile Pasteur pipettes (2)

Demonstration (second period)
Dissected control larvae

Erwinia: Carrot Soft Rot
Petri plate containing nutrient agar

Carrot

Scalpel or razor blade

Potato peeler

Forceps

Alcohol

Disinfectant (bleach)

Sterile water

Sterile Petri plate with filter paper

Cultures
Bacillus: Milky Spore Disease
Bacillus thuringiensis

Erwinia: Carrot Soft Rot
Erwinia carotovora

Techniques Required

Compound light microscopy, Exercise 1

Gram staining, Exercise 5

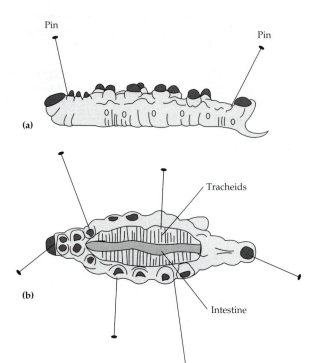

Figure 41.1
Hornworm larva necropsy. **(a)** Dissection. **(b)** Internal anatomy.

Plate streaking, Exercise 11

Starch hydrolysis, Exercise 13

Procedure

Bacillus: Milky Spore Disease

1. Divide each Petri plate into three sectors; label them "A," "B," and "C." Streak a loopful of nutrient broth onto sector A and a loopful of *Bacillus* onto sector B of each plate. Incubate the plates, inverted, at room temperature until growth occurs, then store them in the refrigerator.
2. Make a smear of *Bacillus* and heat-fix it; store it in your drawer.
3. Carefully examine your larvae. Include records of their size, weight, and color.
4. Infect one larva by adding a few drops of *B. thuringiensis* to its food supply. As the infected larva begins to show symptoms of the disease, and as the disease progresses, record the symptoms.
5. Feed the remaining larva in the same manner with nutrient broth. What is the purpose of this step?

    _____

    _____

6. The inoculated larva should die within 1 week. Once it is dead, record its size, weight, and external appearance, place it in a wet towel, and keep it in the refrigerator until you have time to necropsy the larva. *If your larva does not die,* provide some reasons for its survival.
7. Observe the intestine of the dissected control larva.

8. Work with the larva on a disinfectant-saturated towel. Disinfect your instruments by putting them in alcohol and burning off the alcohol immediately before and after use. See color plate VII.4.

> ⚠ **Keep the flame away from the beaker of alcohol.**

Slit the ventral surface open from anus to mouth, then make horizontal cuts to expose the intestine (Figure 41.1). Examine and record your findings. Transfer a loopful of the hemolymph (blood) to sector C on your plates stored previously in the refrigerator (step 1), and incubate them at room temperature for 48 hours. Make a smear of hemolymph on the same slide you used in step 2. Discard the larva as instructed.

> ⚠ **These insects are agricultural pests and cannot be released.**

9. Gram stain the smears and examine the growth on the plates. Growth on the plates can also be Gram stained. Flood the starch plate with Gram's iodine. Record your results.

Erwinia: Carrot Soft Rot*

1. Wash the carrot well, and peel it to eliminate the outer surface; then dry it, wash it with disinfectant, and rinse it with sterile water. Dip your scalpel in alcohol, burn off the alcohol, and cut the carrot into four cross-sectional slices 5 to 8 mm thick.
2. Put the four slices on filter paper in the bottom of a Petri plate.
3. Inoculate the center of three slices with a loopful of the *Erwinia* culture. Why not all four? _____

    _____

    Saturate the filter paper with sterile water. Incubate the plate right-side up at room temperature until soft rot appears. (See color plate VII.3.) More sterile water may be added if the disease process is slow.
4. Divide the nutrient agar plate in half. Inoculate one-half of the nutrient agar with the *Erwinia* broth, incubate the plate, inverted, for 48 hours at room temperature, and then refrigerate it. Make a smear of the *Erwinia.* Heat-fix the smear and store it in your drawer.
5. Streak an inoculum from the diseased carrot on the remaining half of the Petri plate. Incubate the plate, inverted, at room temperature for 48 hours.
6. Make a smear from the diseased carrot. Gram stain both smears and record your observations.
7. Observe the nutrient agar plate and record your results. Prepare a Gram stain from the nutrient agar cultures if time permits.

*Adapted from R. S. Hogue. "Demonstration of Koch's Postulates." *American Biology Teacher* 33:174–175, 1971.

exercise 41

Koch's Postulates

Name _____

Date _____

Lab Section _____

Purpose _____

_____

_____

Bacillus: Milky Spore Disease

Gram Stains

Bacillus thuringiensis

Morphology: _____

Gram reaction: _____

Hemolymph from dead larva

Morphology: _____

Gram reaction: _____

Cultures

Describe the growth on your plates.

Inoculum	Nutrient Agar	Gram Stain	Starch Agar	Starch Hydrolysis?
Broth				
Bacillus culture				
Hemolymph				

Describe the larvae by filling in the following table.

	Larva 1	Larva 2
Initial appearance Date, time		
Length		
Weight		
Color		
Activity		
Which larva is infected?		
After death of infected larva Date, time		
Length		
Weight		
Color		
Appearance of intestine		
Appearance of hemolymph		

Conclusions _____

_____

_____

_____

Questions

1. What evidence would indicate that the microbe obtained from the dead larva was *B. thuringiensis*? _____

_____

_____

_____

2. How would you determine the ID_{50} for *B. thuringiensis*? _____

_____

_____

3. What parts of Koch's postulates did this experiment fulfill? _____

4. What is the normal method of transmission of *Bacillus thuringiensis*? _____

5. *Bacillus thuringiensis* endospores in an inert matrix are sold in garden supply stores. Describe the properties of this insecticide that make it better than chemical insecticides. _____

_____

_____

Erwinia: Carrot Soft Rot

Gram Stains

Erwinia carotovora

Morphology: _____

Gram reaction: _____

Soft Rot

Morphology: _____

Gram reaction: _____

Cultures

Describe the growth on your plate.

Incoculum	Colony Description	Gram Stain
Erwinia culture		
Soft rot		

Describe the appearance of the carrots. _____

_____

_____

_____

Conclusions _____

_____

_____

_____

Questions

1. Why is the disease called soft rot? _____

   _____

2. Why will the addition of water speed the disease process? _____

   _____

   _____

3. What parts of Koch's postulates did this experiment fulfill? _____

4. What is the normal method of transmission of *Erwinia carotovora*? _____

Critical Thinking

1. Eighty-one patients became ill and 14 died in an outbreak at St. Elizabeth's Hospital, Washington, D.C., in 1965. Epidemiologic evidence suggested a link between infection and wind-blown dust from excavations on hospital grounds. In 1968, 144 cases of self-limited illness occurred in employees at the health department building in Pontiac, Michigan. Investigations at that time demonstrated that the etiologic agent was present in the condenser of a malfunctioning air-conditioning system. In 1974, at least 20 persons attending a convention at the Bellevue Stratford Hotel in Philadelphia developed pneumonia, and two died. In 1976, 182 people became ill and 29 died at another convention at the Bellevue Stratford Hotel. What would you need to do to identify the causative agent of these outbreaks?

2. Near Four Corners, New Mexico, on May 14, two people living in the same household died within 5 days of each other. Their illnesses were characterized by flulike symptoms with rapid respiratory failure. The third case was the death of an 8-year-old girl in Mississippi. Over the next 120 days, 53 more people became ill near Four Corners, and 26 died. One man became ill 2 weeks after returning home from a visit to New Mexico. On July 9, a previously healthy woman living in eastern Texas died following acute respiratory distress. The woman had not traveled outside of Texas. What evidence do you have that this is an infectious disease?

part eleven *Immunology*

EXERCISES

In this part, we will examine our defenses against invasion by microorganisms. In 1883, Elie Metchnikoff began his investigations into the body's defenses with this entry in his diary:

> *These wandering cells in the body of the larva of a starfish, these cells eat food, they gobble up carmine granules—but they must eat up microbes too! Of course—the wandering cells are what protect the starfish from microbes! Our wandering cells, the white cells of blood—they must be what protects us from invading germs.* *

These white blood cells (see the photograph on the next page) and certain chemicals are considered agents of nonspecific immunity because they will combat any microorganism that invades the body. Factors involved in nonspecific defenses are the topic of Exercise 42.

The role of specific defenses—or **immunity**—was demonstrated when Emil von Behring developed what he called diphtheria "antitoxin" and, in 1894, successfully treated humans with it. Immunity involves the production by the body of specific proteins called **antibodies,** which are directed against distinct microorganisms. If microorganism A invades the body, antibodies are produced against A; if microorganism B invades the

*Quoted in P. DeKruif. *Microbe Hunters*. New York: Harcourt, Brace, & World, 1953, p. 195.

body, antibodies are produced against B. The antibodies produced against microorganism B will not react with microorganism A. The chemical substance that induces antibody production is called an **antigen.** In the examples just mentioned, microorganisms A and B possess antigens on their surfaces.

Specific and nonspecific immunity involves blood. **Blood** consists of a fluid called **plasma** and formed elements: cells and cell fragments. The cells that are of immunologic importance are the white blood cells. **Serum** is the fluid portion that remains after blood has clotted. We will study the techniques and mechanisms of **serology,** or antigen–antibody reactions in vitro, in Exercises 43 through 45.

Early stage of phagocytosis by an alveolar macrophage (a type of macrophage found in the alveoli of the lungs). The "wrinkled" macrophage has contacted and adhered to a smooth, roughly spherical yeast cell by means of a pseudopod (3100×).

exercise 42

Nonspecific Resistance

Objectives

After completing this exercise, you should be able to:

1. Differentiate between specific and nonspecific immunity.
2. List and discuss the functions of at least two chemicals involved in nonspecific resistance.
3. Use a spectrophotometer.
4. Determine the effects of normal serum bactericidins.

Background

Our ability to ward off diseases through our body's defenses is called **resistance.** Resistance can be divided into two kinds: specific and nonspecific. Specific resistance, or **immunity,** is the defense against a specific microorganism. **Nonspecific resistance** refers to all our defenses that protect us from invasion by any microorganism. Physical barriers, such as the skin and mucous membranes, are part of nonspecific resistance. Inflammation and phagocytosis play major roles in nonspecific resistance, as do chemicals produced by the body. As early as 1888, Metchnikoff noted that bacteria "show degenerative changes" when inoculated into samples of mammalian blood.

Lysozyme and complement are examples of chemicals involved in nonspecific resistance. These proteins are collectively called **bactericidins. Lysozyme** is an enzyme found in body fluids that is capable of breaking down the cell walls of gram-positive bacteria and a few gram-negative bacteria. **Complement** is a group of proteins found in normal serum that are involved in phagocytosis and lysis of bacteria. Complement can be activated by bacterial cell wall polysaccharides and antigen–antibody reactions.

In this exercise, we will examine chemicals involved in nonspecific resistance.

Materials

Lysozyme Activity

Lysozyme buffer, 4.5 ml

Egg-white lysozyme, 1:10⁵ dilution, 2.5 ml

Spectrophotometer tubes (2)

1 ml pipette

5 ml pipette (5)

Petri dish

Spectrophotometer

Normal Serum Bactericidins

Nutrient agar, melted and cooled to 45°C

Sterile Petri dishes (9)

Sterile 1 ml pipettes (10)

Sterile 99 ml dilution blanks (3)

Sterile serological tubes (9)

Serological test tube rack

0.85% saline solution

Normal animal serum

Animal serum heated to 56°C for 30 minutes

Cultures (as assigned)

Micrococcus lysodeikticus suspension, 5 ml

Escherichia coli

Staphylococcus aureus

Techniques Required

Inoculating loop, Exercise 10

Pour plate, Exercise 11

Pipetting, Appendix A

Serial dilution technique, Appendix B

Spectrophotometry, Appendix C

Graphing, Appendix D

Procedure

 Work only with your own tears or saliva.

Lysozyme Activity

1. Collect tears or saliva in a Petri dish, as explained by your instructor.
2. Prepare a 1:10 dilution of tears or saliva by adding 0.5 ml tears or saliva to 4.5 ml lysozyme buffer. What is the purpose of the buffer? _____

3. Mix 2.5 ml of the tear or saliva preparation with a 2.5 ml *Micrococcus lysodeikticus* suspension in a spectrophotometer tube. Carefully pipette the contents of the tube up and down three times to mix.

4. Record the absorbance at 540 nm at 30 seconds, 60 seconds, 120 seconds, 180 seconds, 240 seconds, 5 minutes, and at 5-minute intervals for the next 15 minutes.

5. Repeat steps 3 and 4 using egg-white lysozyme.

6. Plot your data in the Laboratory Report.

 Place the Petri plate, dilution tube, and spectrophotometer tube in disinfectant.

Normal Serum Bactericidins (Figure 42.1)

Work in groups as assigned.

1. Aseptically prepare 1:10^2, 1:10^4, and 1:10^6 dilutions of the culture (*S. aureus* or *E. coli*) assigned to you (Figure 42.1a).

2. Label three sterile serological tubes "6A," "6B," and "6C." Aseptically transfer 0.1 ml of culture from the highest dilution (1: _____) to each tube (Figure 42.1b).

3. Repeat step 2 with the other dilutions using the same pipette. Label each tube appropriately. How many tubes do you have? _____

4. To each A tube, add 0.4 ml of unheated normal serum; to each B tube, add 0.4 ml of serum heated to 56°C for 30 minutes; to each C tube, add 0.4 ml of saline (Figure 42.1c, d, and e). What is the purpose of the C tubes? _____

5. Mix all tubes and incubate them at 35°C for 1 hour (Figure 42.1f).

6. Label nine sterile Petri dishes in the same manner as the tubes: "1:10^2 A," "1:10^2 B," "1:10^2 C," "1:10^4 A," "1:10^4 B," and so on.

7. Remove 0.1 ml from the A tube of the highest dilution, and place it in the appropriate Petri dish. Using the same pipette, transfer 0.1 ml from the 1:10^4 A tube to the corresponding Petri dish. Repeat this procedure with the 1:10^2 A tube (Figure 42.1g). Why are the samples transferred from the highest dilution first? _____

8. Using another pipette, repeat step 7 with the B tubes. With your remaining pipette, repeat step 7 with the C tubes.

9. Pour melted, cooled nutrient agar into each plate to a depth of approximately 5 mm. Gently swirl to mix the contents, and allow the agar to solidify (Figure 42.1h).

10. Incubate the plates for 24 to 48 hours at 35°C. Record the number of colonies on each plate. Plates with fewer than 25 colonies should be reported as TFTC (too few to count), and those with more than 250 colonies as TNTC (too numerous to count).

11. Calculate the number of colony-forming units (cfu) per milliliter for the control. Choose one plate with between 25 and 250 colonies:

$$\frac{\text{Number of colonies}}{0.1 \text{ ml} \times \text{dilution}} = \text{cfu/ml}$$

Calculate the number of colony-forming units per milliliter for the unheated and heated sera.

12. Calculate the percent of increase or decrease in bacterial numbers due to exposure to heated and unheated serum as compared to the control.

$$\frac{\text{Control} - \text{Experiment}}{\text{Control}} \times 100 = \%$$

13. Compare your data with those of a group using the other bacterial species.

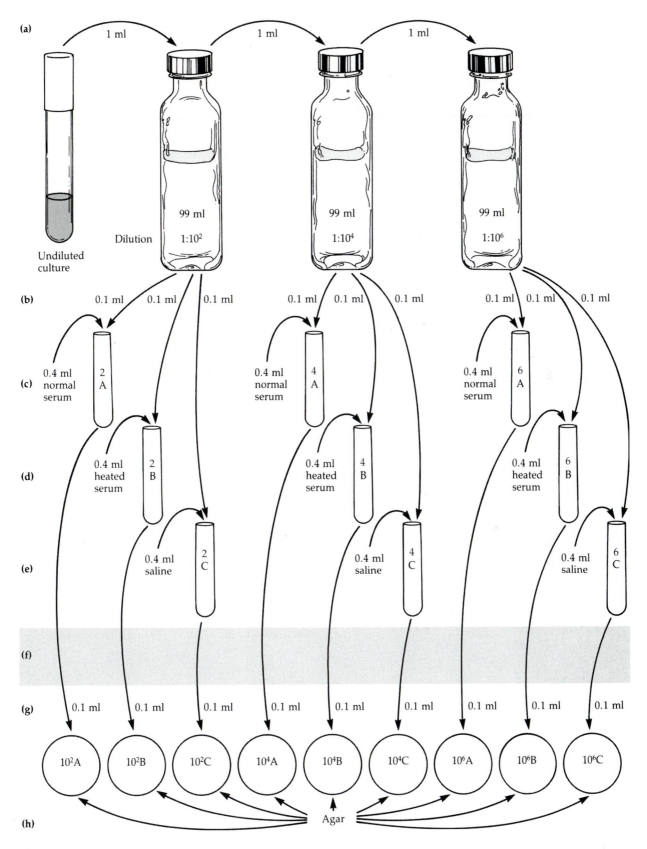

Figure 42.1
Determining the bactericidal activity of serum.

LABORATORY REPORT

Nonspecific Resistance

Name _____

Date _____

Lab Section _____

Purpose _____

_____

_____

Data

Lysozyme Activity

Time	Absorbance		
	Saliva	Tears	Egg-White Lysozymes
30 sec			
60 sec			
120 sec			
180 sec			
240 sec			
5 min			
10 min			
15 min			
20 min			

Plot your data for lysozyme activity on the graph paper on the next page. Absorbance is marked on the Y-axis and time on the X-axis. Make one line for tears or saliva and another line for egg-white lysozyme.

Normal Serum Bactericidins

	Number of Colonies			cfu/ml
	$1:10^2$	$1:10^4$	$1:10^6$	
Unheated serum				
Heated serum				
Control				

Your calculations:

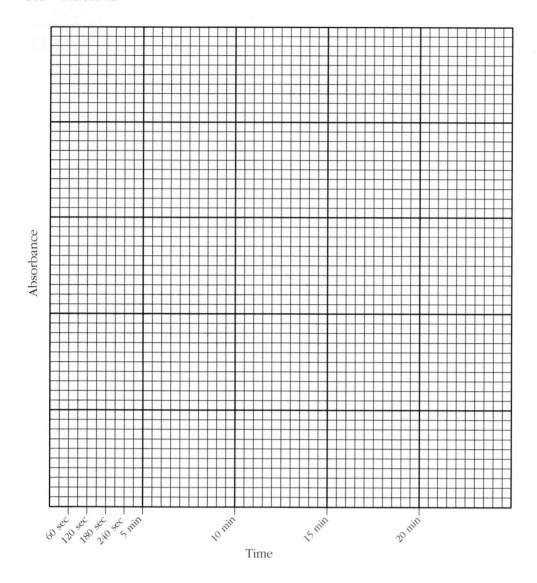

Conclusions

What effect did lysozyme have on the bacterial suspension? _____

_____

	% Increase or Decrease	
	S. aureus	*E. coli*
Unheated serum		
Heated serum		

Your calculations:

Questions

1. Why are tears and saliva potential biohazards? _____

    _____

2. Was lysozyme present in your tears or saliva? _____

    How can you tell? _____

3. Did you test tears or saliva? _____ Compare your data with someone who tested

    the other. Which secretion has more lysozyme activity? _____ How can you tell?

    _____

4. Does normal serum have bactericidal properties? Explain. _____

    _____

Critical Thinking

1. What other factors contribute to the nonspecific resistance of your eyes? Of your mouth?

2. What can you conclude regarding the effect of heat on the bactericidal properties of normal serum?

3. Design an experiment to determine whether plant fluids have bactericidal properties.

exercise 43

Blood Group Determination: Slide Agglutination

Objectives

After completing this exercise, you should be able to:

1. Compare and contrast the terms agglutination and hemagglutination.
2. Determine ABO and Rh blood types.
3. Determine possible compatible transfusions.

Background

Agglutination reactions occur between high-molecular-weight, particulate antigens and antibodies. Since many antigens are on cells, agglutination reactions lead to the clumping, or **agglutination,** of cells. When the cells involved are red blood cells, the reaction is called **hemagglutination.**

Hemagglutination reactions are used in the typing of blood. The presence or absence of two very similar carbohydrate antigens (designated A and B) located on the surface of red blood cells is determined using specific antisera (Figure 43.1). Hemagglutination occurs when anti-A antiserum is mixed with type A red blood cells. When anti-A antiserum is mixed with type B red blood cells, no hemagglutination occurs. People with type AB blood possess both A and B antigens on their red blood cells, and those with type O blood lack A and B antigens (see Figure 43.1).

Many other blood antigen series exist on human red blood cells. Another surface antigen on red blood cells is designated the Rh factor. The **Rh factor** is a complex of many antigens. The Rh factor that is used routinely in blood typing is the **Rh_0 antigen,** or **D antigen.** Individuals are Rh-positive when D antigen is present. The presence of the Rh factor is determined by a hemagglutination reaction between anti-D antiserum and red blood cells with D antigen on their surface.

A person possesses antibodies to the alternate antigen. Thus, people of blood type A will have antibodies to the B antigen in their sera (see Figure 43.1). Rh-negative individuals do not naturally have anti-D antibodies in their sera. Anti-D antibodies are produced when red blood cells with D antigen are introduced into Rh-negative individuals.

The ABO and Rh systems place restrictions on how blood may be transfused from one person to another. An incompatible transfusion results when the antigens of the donor red blood cells react with the antibodies in the recipient's serum or induce the formation of antibodies.

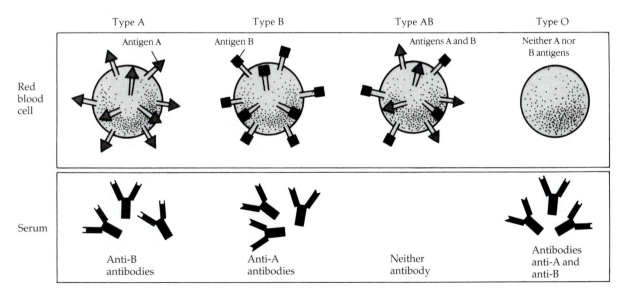

Figure 43.1

Relationship of antigens and antibodies involved in the ABO blood group system.

Table 43.1

The ABO and Rh Blood Group Systems

Characteristic	Blood Type					
	A	B	AB	O	Rh+	Rh−
Antigen present on the red blood cells	A	B	Both A and B	Neither A nor B	D	No D
Antibody normally present in the serum	Anti-B	Anti-A	Neither anti-A nor anti-B	Both anti-A and anti-B	No anti-D	No anti-D*
Serum causes agglutination of red blood cells of these types	B, AB	A, AB	None	A, B, AB	Neither Rh+ nor Rh−	Neither Rh+ nor Rh−*
Percent occurrence in a mixed Caucasian population	41	10	4	45	85	15
Percent occurrence in a mixed Black population	27	20	7	46	85	15

*Anti-D antibodies are not naturally present in the serum of Rh− people. Anti-D antibodies can be produced upon exposure to the D antigen through blood transfusions or pregnancy.

Source: Adapted from G. J. Tortora, B. R. Funke, and C. L. Case. *Microbiology: An Introduction*, 7th ed. San Francisco, CA: Benjamin Cummings, 2001.

A summary of the major characteristics of the ABO and Rh blood groups is presented in Table 43.1.

Materials

Cotton moistened with 70% ethyl alcohol

Sterile cotton balls and bandage

Sterile lancet

Anti-A, anti-B, and anti-D antisera

Glass slides (2)

Toothpicks

Grease pencil

Techniques Required

Dissecting microscope, Appendix E

 Carefully read the Safety Precautions for Working with Blood in the Laboratory.

Procedure

1. With a grease pencil, draw two circles on a clean glass slide, and label one "A" and the other "B." Draw a circle on the second slide and label it "D."

 Safety Precautions for Working with Blood in the Laboratory

- Do not perform this lab experiment if you are sick.

- Work only with **your own** blood* or wear gloves.

- **Disinfect** your finger with 70% alcohol. Use any finger except your thumb. Unwrap a **new, sterile** lancet and pierce the disinfected finger with the sterile lancet.

- Place the used lancet **in disinfectant** in a container designated by the instructor or a biohazard "sharps" container.

- Stop the bleeding with a sterile cotton ball and **apply an adhesive bandage.**

- **Dispose of** the used cotton **in disinfectant.**

- **Discard** used slides and toothpicks **in disinfectant.**

- **Wash** any spilled blood from your work area **with disinfectant.**

*Blood obtained from blood banks is tested for hepatitis B virus and HIV. However, no test method can offer complete assurance that laboratory specimens do not contain these viruses.

2. Disinfect your middle finger with cotton saturated in alcohol (Figure 43.2). Pierce the disinfected finger with a sterile lancet.

3. Let a drop of blood fall into each circle. Stop the bleeding with a sterile cotton ball, and apply an adhesive bandage.

4. To the A circle, add 1 drop of anti-A antiserum. Add 1 drop of anti-B antiserum to the B circle and 1 drop of anti-D antiserum to the D circle.

5. Mix each suspension with toothpicks. Use a different toothpick for each antiserum. Why? _____

_____

6. Observe for agglutination, and determine your blood type. (See color plate IX.1.) A dissecting microscope may help you see hemagglutination.

> ☣ **Discard the slides, toothpicks, cotton balls, and lancet in the disinfectant.**

Wipe in a circular pattern

Alcohol-soaked cotton

Figure 43.2
To disinfect skin, rub it with 70% ethyl alcohol.

exercise 43   **LABORATORY REPORT**

Blood Group Determination: Slide Agglutination

Name _____

Date _____

Lab Section _____

Purpose _____

_____

_____

Data

Antiserum	Hemagglutination
Anti-A	
Anti-B	
Anti-D	

Data Analysis

On the blackboard, tabulate the blood types for the class or laboratory section. Calculate the percentage of each blood type, and compare with the percent distribution given in Table 43.1.

Blood Type	Number of Students	Percent
A		
B		
AB		
O		

Questions

1. What is your blood type? _____

2. What is the blood type of the person on your left? _____

   Is your blood potentially compatible? _____ Explain briefly. _____

3. What is the blood type of the person on your right? _____

Is your blood potentially compatible? _____ Explain briefly. _____

Critical Thinking

1. What is hemolytic disease of the newborn? How does Rhogam® prevent it?

2. How can blood type O be considered the "universal donor"?

3. Individuals have antibodies to the AB blood antigens not found on their red blood cells. What is the origin of these antibodies in someone who has never had a transfusion?

4. A woman with type A blood can have a healthy baby with type B blood. Why doesn't the baby develop hemolytic disease of the newborn?

exercise 44

Agglutination Reactions: Microtiter Agglutination

Objectives

After completing this exercise, you should be able to:

1. Define agglutination and titer.
2. Determine the titer of antibodies by the agglutination method.
3. Provide two applications for agglutination reactions.

Background

The outer surfaces of bacterial cells contain antigens that can be used in agglutination reactions. **Agglutination reactions** occur between *particulate* antigens—such as cell walls, flagella, or capsules *bound* to cells—and antibodies. **Agglutination,** or the clumping of bacteria by antibodies, is a useful laboratory diagnostic technique.

In a slide agglutination test, an unknown bacterium is suspended in a saline solution on a slide and mixed with a drop of known antiserum. This test is done with different antisera on separate slides. The bacteria will agglutinate when mixed with antibodies produced against the same strain and species. A positive test can identify the bacterium.

An **agglutination titration** can estimate the concentration (*titer*) of antibody in serum. This test is used to determine whether a particular organism is causing the patient's symptoms. If an increase in titer is shown in successive daily tests, the patient probably has an infection caused by the organism used in the test.

In the titration, bacteria are mixed with dilutions of serum (e.g., 1:10, 1:20, 1:40, and so on). The *endpoint* is the greatest dilution of serum showing an agglutination reaction. The *reciprocal* of this dilution is the *titer.* For example, in Figure 44.1a, the titer is 80.

In this exercise, we will perform a test to diagnose typhoid fever, the **Widal test.**

| 1:10 | 1:20 | 1:40 | 1:80 | 1:160 | 1:320 | Control |

(a) Top view of wells

Agglutination

(b) Enlarged side view of wells

(c)

Figure 44.1

(a) Each well in this microtiter plate contains, from left to right, only half the concentration of serum in the preceding well. Each well contains the same concentration of bacterial cells. **(b)** In a positive reaction, sufficient antibodies are present in the serum to link the antigens together, forming an antibody–antigen mat that sinks to the bottom of the well. In a negative reaction, insufficient antibodies are present to cause the linking of antigens. **(c)** Appearance of wells after refrigeration for 24 hours. The first two wells show agglutination. The endpoint is 1:20.

Figure 44.2

Dilution of antiserum. Twenty μl of patient's serum is mixed with 80 μl of saline in well 1. Serial dilutions are made by transferring 50 μl from one well to the next, through well 9. Discard 50 μl from well 9.

 ***Salmonella typhimurium* is a causative agent of gastroenteritis. Be careful when handling this pathogenic bacterial species.**

Materials

0.85% saline solution

"Patient's" serum

Microtitration plate

Micropipettes, 20 to 80 μl

Micropipette tips (13)

Culture

Salmonella typhimurium (heated for 30 minutes at 56°C in water bath)

Techniques Required

Pipetting, Appendix A

Serial dilution technique, Appendix B

Procedure

1. Label ten wells in the microtiter plate "1" through "10" (Figure 44.2).
2. Add 80 μl saline to the first well and 50 μl to each of the remaining wells.
3. Add 20 μl patient's serum to the first well. Mix up and down three times and, with a new pipette tip, transfer 50 μl to the second well. Change the pipette tip, mix, and transfer 50 μl to the third well; and so on. Continue until you have reached the ninth well. Discard 50 μl from that well, as shown in Figure 44.2. The patient's serum has now been diluted. The dilutions are as follows:

Well 1   1:5

Well 2   1:10

Well 3   1:20

Well 4   1: _____

Well 5   1: _____

Well 6   1: _____

Well 7   1: _____

Well 8   1: _____

Well 9   1: _____

Well 10   no serum

Fill in the missing dilutions.

4. Carefully add 50 μl of the antigen to each well. What is the antigen? _____
   How does the addition of 50 μl of antigen affect the dilutions? _____
5. Shake the plate carefully in a horizontal direction to mix the contents in each well. Place the plate in a 35°C incubator for 60 minutes.
6. Refrigerate the plate until the next laboratory period.
7. Observe the bottom of each well for agglutination. A dissecting microscope may help you to see agglutination. Which well serves as the control? _____

   _____

   What should occur in this well? _____

   _____

Determine the endpoint and the titer.

exercise 44

Agglutination Reactions: Microtiter Agglutination

Name _____

Date _____

Lab Section _____

Purpose _____

_____

_____

Data

Well #	Final Dilution	Agglutination
1		
2		
3		
4		
5		
6		
7		
8		
9		
10		

Diagram the appearance of the bottom of a positive and a negative well:

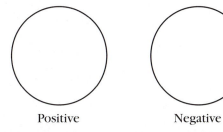

Positive   Negative

Questions

1. How did you determine the presence of agglutination? _____

    _____

2. What was the endpoint? _____

3. What was the antibody titer? _____

4. What was the antiserum used in this experiment? _____

Critical Thinking

1. How would agglutination reactions help in locating the source of an epidemic?

2. The following antibody titers were obtained for three patients:

Patient	Antibody Titer		
	Day 1	Day 5	Day 12
A	128	128	128
B	128	256	512
C	0	0	0

    What can you conclude about each of these patients?

3. An agglutination test was performed on a patient's blood. The titer was 512, and upon further testing the antibody involved was identified as IgM. Can any conclusions be drawn?

exercise 45

ELISA Technique

"A scientist is like a learned child. At some point, though, children have to put away their things and stop their playing. But a scientist can stay that way for a lifetime."

"Scientific discovery is a bit like digging for gold. The pioneering work gets done by individualists, frequently following lonesome trails, commonly a little eccentric, and often failing in their quests. Every now and then an inspired or lucky strike turns up a vein rich and promising enough to draw a crowd."

GEORGE WALD

Objectives

After completing this exercise, you should be able to:

1. Explain how ELISA tests can be used clinically to detect antibodies or antigens.
2. Differentiate between direct and indirect ELISA tests.
3. Determine the titer of antibodies by the ELISA method.

Background

Diseases can be diagnosed by using known antibodies to detect the presence of an antigen or by detecting specific antibodies in a patient. These types of tests are called **immunoassays.** Immunoassays are based on detectable interactions between antigens and antibodies such as precipitation, agglutination, or complement fixation. Increased sensitivity in detecting antigens can be achieved by labeling antibodies with substances such as radioactive chemicals (e.g., Iodine-125), fluorescent compounds, magnetic beads, or enzymes. **ELISA (Enzyme-Linked Immunosorbent Assay)** or **EIA (Enzyme Immunosorbent Assay)** is the most widely used immunoassay in labs today. ELISAs use enzyme labels because of their stability, reproducibility, safety, ease of detection, and relatively low cost. The enzymes horseradish peroxidase and alkaline phosphatase are most often used. When the appropriate substrate is added, the enzyme reacts with the substrate to make a colored product, which can be detected visually or by a spectrophotometer. The amount of product produced is directly proportional to the amount of enzyme—and, therefore, antibody—present, allowing the technique to be quantitative as well as qualitative.

In the *direct* ELISA technique, enzyme-labeled antibodies are used to identify an antigen. In the direct ELISA technique, a known antibody is adsorbed to the wells in a microtiter plate. The unknown microorganism from a patient is added to the wells. If the antibody in the well is specific for the microorganism, the microbe will be bound to the antibody. A second antibody specific for the antigen is then added. This antibody is linked to an enzyme. The reaction between the antigen and antibody is made visible by addition of the substrate for the enzyme (Figure 45.1a).

Indirect ELISAs are used to detect a specific antibody in a patient's serum. In the indirect technique, the known antigen is attached to the wells of a microtiter plate. Dilutions of the suspected antibody are added and, after washing, enzyme-labeled anti-antibody is added. A color change after adding the substrate indicates that the antibody was present (Figure 45.1b). The ELISA test for HIV antibodies is one example of the indirect technique.

In this exercise, we will perform an ELISA test to diagnose salmonellosis.

 ***Salmonella typhimurium* is a causative agent of gastroenteritis. Be careful when handling this pathogenic bacterial species.**

Materials

Coating buffer

Washing buffer

Blocking buffer

Patient's serum

Alkaline phosphatase–labeled anti-antibodies

BCIP/NBT substrate

Polystyrene flat-bottom microtiter plate

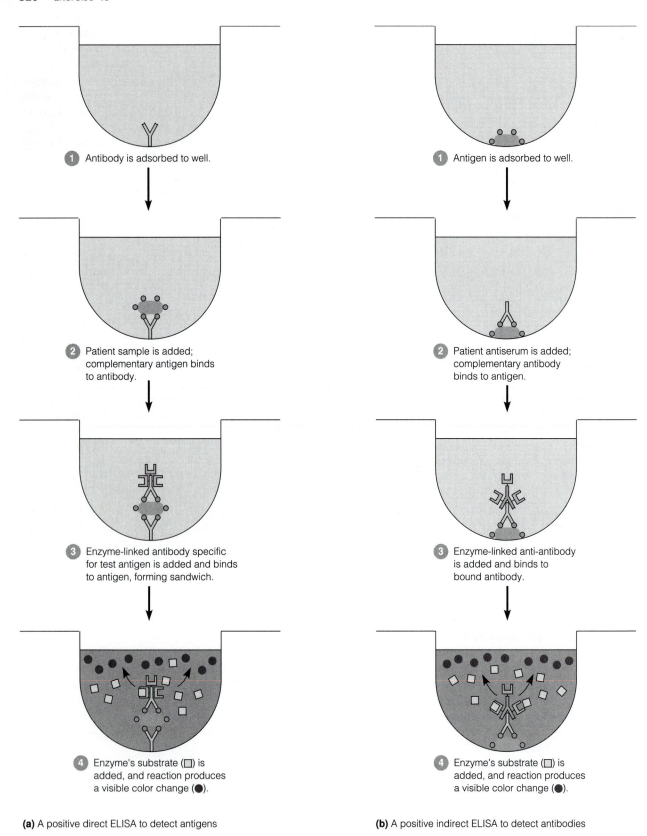

1 Antibody is adsorbed to well.

2 Patient sample is added; complementary antigen binds to antibody.

3 Enzyme-linked antibody specific for test antigen is added and binds to antigen, forming sandwich.

4 Enzyme's substrate (☐) is added, and reaction produces a visible color change (●).

(a) A positive direct ELISA to detect antigens

1 Antigen is adsorbed to well.

2 Patient antiserum is added; complementary antibody binds to antigen.

3 Enzyme-linked anti-antibody is added and binds to bound antibody.

4 Enzyme's substrate (☐) is added, and reaction produces a visible color change (●).

(b) A positive indirect ELISA to detect antibodies

Figure 45.1
The ELISA method. The components are usually contained in small wells of a microtiter plate.

Clear plastic tape

Micropipettes, 50 to 100 μl

Micropipette tips

Culture

Salmonella typhimurium (heated for 30 minutes at 56°C in a water bath)

Techniques Required

Pipetting, Appendix A

Serial dilution technique, Appendix B

Microtiter dilutions, Exercise 44

Procedure

1. Add 100 μl coating buffer to each well of one row (wells 1 – 12) of the microtiter plate.
2. Add 100 μl *S. typhimurium* to each well.
3. Seal the wells with a strip of plastic tape, and refrigerate the plate at 5°C for 1 to 7 days.
4. Remove your plate from the refrigerator and carefully remove the tape. Shake the inverted plate with a quick shake to remove the liquid into the disinfectant.
5. Fill the wells with washing buffer and shake to remove. Wash two more times.
6. Add 100 μl blocking buffer. Leave for 30 to 90 minutes as directed by your instructor.

7. Perform dilutions of the patient's serum by placing 100 μl in the first well. Mix up and down three times and, with a new pipette tip, transfer 100 μl to the second well. Mix up and down three times, change pipette tips, and transfer 100 μl to the third well, and so on. Continue until you have reached the 11th well. Discard 100 μl from that well. The patient's serum has now been diluted from 1:2 to 1:_____ in the 11th well. What is in well 12? _____

   What is the purpose of well 12? _____

   _____

   _____

8. Incubate the plate at 35°C for 60 minutes.
9. Shake the inverted plate with a quick shake to remove the contents. Wash three times with washing buffer as described in step 5.
10. Add 100 μl of alkaline phosphatase – labeled anti-antibody to each well (1 – 12). Seal the wells with tape and incubate the plate at 35°C for 45 minutes. Plates can be sealed and stored at 5°C until the next lab period. What is the antigen? ___

   _____

11. Remove the tape carefully, shake out the contents, and wash the wells three times with washing buffer.
12. Add 100 μl of the alkaline phosphatase substrate (BCIP/NBT) to each well in the row.
13. Leave at room temperature for 10 to 30 minutes until color develops; well 12 will be colorless.
14. Record the results. The highest dilution with a blue color is the endpoint. The titer is the reciprocal of the dilution of the endpoint.

exercise 45

ELISA Technique

Name _____

Date _____

Lab Section _____

Purpose _____

_____

_____

Data

Well #	Final Dilution	Color
1		
2		
3		
4		
5		
6		
7		
8		
9		
10		
11		
12		

Questions

1. What was the endpoint? _____

2. What was the antibody titer? _____

3. Was the test performed in this exercise a direct or an indirect ELISA test? _____

4. Why was the control well colorless? _____

    _____

    _____

Critical Thinking

1. Compare and contrast direct and indirect ELISA techniques.

2. What is the purpose of anti-antibodies in the ELISA technique?

3. Describe how you would determine if a bacterium isolated from a patient was *Salmonella* using the ELISA technique with a known anti-*Salmonella* antiserum.

4. Why would the ELISA technique be valuable for identification of viruses?

5. ELISAs are commonly used in clinical laboratories. What advantage does the ELISA technique have over fluorescent antibody or radioactive antibody techniques?

part twelve

Microorganisms and Disease

EXERCISES

Many microorganisms grow abundantly both inside and on the surface of the normal adult body. The microorganisms that establish more or less permanent residence without producing diseases are known as **normal microbiota** (see the photograph on the next page). Microorganisms that may be present for a few days or months are called **transient microbiota.**

The relationship between normal microbiota and a healthy person may be classified as one of two types of **symbiosis** (the close association between two organisms): **commensalism** or **mutualism.** In a *commensal* relationship, one organism is benefited and the other is unharmed. For example, many of the bacteria living in the large intestine are supplied with continual food and a constant temperature, while the host is neither benefited nor harmed. In a *mutualistic* relationship, both organisms are benefited. Some bacteria that live on and in our bodies receive a constant supply of nutrients, while they, in exchange, supply something of benefit to us. For example, *E. coli* synthesizes vitamin K and certain B vitamins used by its human host.

A third kind of symbiosis is **parasitism.** In this relationship, one organism gains while the other is harmed. In a disease state, the microorganism is a parasite and is harming its human host. An organism like this is called a **pathogen.**

Lactic acid bacteria on the surface of the stomach epithelium.

The pathogens that cause disease consist of many different organisms. Robert Koch speculated in one of his early publications:

> *On this . . . I take my stand, and, till the cultivation of bacteria from spore to spore shows that I am wrong, I shall look on pathogenic bacteria as consisting of different species . . . .*

Some members of the normal microbiota become parasites under certain conditions. These organisms are called **opportunists.**

In order to cause a disease, a pathogen must gain access to the human body. The route of access is called the **portal of entry.** Some microbes cause disease only when they gain access by the "correct" portal of entry.

In Exercises 46 through 50, we will examine the variety of bacteria associated with the human body. In a clinical laboratory, samples from diseased tissue are cultured, and pathogens must be distinguished from normal and transient microbiota. In Exercise 51, we will identify bacteria in a simulated clinical sample.

Learning with Technology

BID = Bacteria ID CD-ROM

An activity using the Bacteria ID CD-ROM is included in Exercises 46 through 50. This CD-ROM is included free with each new copy of Tortora, Funke, and Case, *Microbiology: An Introduction,* Seventh Edition. If you are not using this text but would like to buy a stand-alone version of this software, you can ask your bookstore to order it for you: ISBN 0-8053-2157-8. To get started using this CD, read the instructions in Appendix G of *Microbiology: An Introduction* (the textbook).

Use these **BID** activities in Exercises 46 through 50 to practice using biochemical tests and the keys in Appendix H and *Bergey's Manual.* This simulation program cannot replace the skills you will gain doing the actual lab work, but it will familiarize you with the organization of bacterial identification schemes and the use of specific tests for distinguishing bacteria. The bacteria to be identified are coded by set and unknown numbers that correspond to those used in the Bacteria ID CD-ROM. When you log in, click on the Tortora level.

*Quoted in T. D. Brock, ed. *Milestones in Microbiology.* Washington, DC: American Society of Microbiology, 1961, p. 99.

exercise 46

Bacteria of the Skin

EXERCISE
46

Objectives

After completing this exercise, you should be able to:

1. Isolate and identify bacteria from the human skin.
2. Provide an example of normal skin microbiota.
3. List characteristics used to identify the staphylococci.
4. Explain why many bacteria are unable to grow on human skin.

Background

The skin is generally an inhospitable environment for most microorganisms. The dry layers of keratin-containing cells that make up the epidermis (the outermost layer of the skin) are not easily colonized by most microbes. Sebum, secreted by oil glands, inhibits bacterial growth, and salts in perspiration create a hypertonic environment. Perspiration and sebum are nutritive for certain microorganisms, however, which establishes them as part of the normal microbiota of the skin.

Normal microbiota of the skin tend to be resistant to drying and to relatively high salt concentrations. More bacteria are found in moist areas, such as the axilla (armpit) and the sides of the nose, than on the dry surfaces of arms or legs. Transient microbiota are present on hands and arms in contact with the environment.

Propionibacterium live in hair follicles on sebum from oil glands. The propionic acid they produce maintains the pH of the skin between 3 and 5, which suppresses the growth of other bacteria. Most bacteria on the skin are gram-positive and salt-tolerant.

Staphylococcus aureus is part of the normal microbiota of the skin and is also considered a pathogen. *S. aureus*, which produces **coagulase,** an enzyme that coagulates (clots) the fibrin in blood, is pathogenic. A test for the presence of coagulase is used to distinguish *S. aureus* from other species of *Staphylococcus*.

Although many different bacterial genera live on human skin, in this exercise we will attempt to isolate and identify a catalase-positive, gram-positive coccus.

Materials

Petri plate containing mannitol salt agar

Sterile cotton swab

Sterile saline

Second period:

Petri plate containing mannitol salt agar

3% hydrogen peroxide

Gram staining reagents

Fermentation tubes

Coagulase plasma (as needed)

Toothpick

Techniques Required

Gram staining, Exercise 5

Plate streaking, Exercise 11

Selective media, Exercise 12

Fermentation tests, Exercise 14

Catalase test, Exercise 17

Procedure

1. Wet the swab with saline, and push the swab against the wall of the test tube to express excess saline (Figure 46.1a). Swab any surface of your skin. The sides of the nose, axilla, an elbow, or a pus-filled sore are possible areas.
2. Swab one-half of the plate with the swab.

 Discard the swab in disinfectant.

Using a sterile loop, streak back and forth into the swabbed area a few times, then streak away from the inoculum (Figure 46.1b).

3. Incubate the plate, inverted, at 35°C for 24 to 48 hours.
4. Examine the colonies. Record the appearance of the colonies and any mannitol fermentation (yellow halos). (See color plate V.2.) Gram stain the colonies and test for catalase production. Perform the catalase test by making a suspension of the desired colony on a slide with a toothpick and adding a drop of H_2O_2. Discard the toothpick in the biohazard container.
5. Subculture a catalase-positive, gram-positive coccus on another mannitol salt plate. Do not attempt

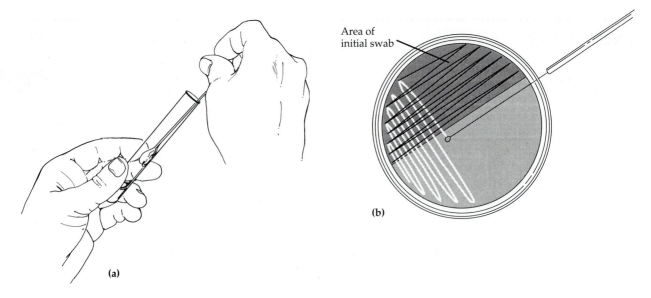

Figure 46.1

Taking a sample from the skin. **(a)** Aseptically moisten a sterile cotton swab in saline. **(b)** After swabbing half the plate, use a sterile loop to streak the inoculum over the remaining half of the plate.

to subculture a colony that has not been tested for catalase. Why?_____

6. Using the key in Appendix H, Figure H.3, proceed to identify your isolate.

a. To test for coagulase, place a loopful of rehydrated coagulase plasma on a clean slide. Add a loopful of water and make a heavy suspension of the bacteria to be tested. Observe for clumping of the bacterial cells (clumping = coagulase-positive; no clumping = coagulase-negative). (See color plate VIII.3.)

b. Inoculate the appropriate fermentation tubes. Incubate the fermentation tubes at 35°C for 24 to 48 hours.

exercise 46

Bacteria of the Skin

Name _____

Date _____

Lab Section _____

Purpose _____

_____

_____

Data

Source of inoculum: _____

First Mannitol Salt Plate

	Colony		
	1	2	3
Colony description			
Pigment			
Mannitol fermentation			

Which colony was isolated? _____

Catalase reaction: _____

Gram stain

Reaction: _____

Morphology: _____

Arrangement: _____

Additional Biochemical Tests

_____

_____

_____

Conclusion

What organism did you identify? _____

Questions

1. Why is mannitol salt agar used as a selective medium for normal skin microbiota? _____

    _____

    _____

2. After you have observed a gram-positive coccus, what is the additional information you need before performing

    a coagulase test? _____

    _____

3. List three identifying characteristics of *Staphylococcus aureus*. _____

    _____

    _____

4. List three factors that protect the skin from infection. _____

    _____

    _____

    _____

Critical Thinking

1. What is coagulase? How is it related to pathogenicity?

2. Assume that you isolated *S. aureus* from your skin. How would you determine whether it was penicillin-resistant?

Learning with Technology

BID Fill in the following table for the organism described in Set 1, Unknown #51. Use the keys in Appendix H to identify the bacterial species.

	Results
Colony description	
Gram reaction and morphology	
Glucose fermentation	
Mannitol fermentation	
Trehalose fermentation	
Catalase reaction	

The bacterium is _____

exercise 47

Bacteria of the Respiratory Tract

Objectives

After completing this exercise, you should be able to:

1. List representative normal microbiota of the respiratory tract.
2. Differentiate the pathogenic streptococci based on biochemical testing.
3. List a characteristic used to identify *Neisseria, Corynebacterium, Mycobacterium,* and *Bordetella.*

Background

The respiratory tract can be divided into two systems: the upper and lower respiratory systems. The **upper respiratory system** consists of the nose and throat, and the **lower respiratory system** consists of the larynx, trachea, bronchial tubes, and alveoli. The lower respiratory tract is normally sterile because of the efficient functioning of the ciliary escalator. The upper respiratory system is in contact with the air we breathe—air contaminated with microorganisms.

The throat is a moist, warm environment, allowing many bacteria to establish residence. Species of many different genera—such as *Staphylococcus, Streptococcus, Neisseria,* and *Haemophilus*—can be found living as normal microbiota in the throat. Despite the presence of potentially pathogenic bacteria in the upper respiratory system, the rate of infection is minimized by microbial **antagonism.** Certain microorganisms of the normal microbiota suppress the growth of other microorganisms through competition for nutrients and production of inhibitory substances.

Streptococcal species are the predominant organisms in throat cultures, and some species are the major cause of bacterial sore throats (acute pharyngitis). Streptococci are identified by biochemical characteristics, including hemolytic reactions, and antigenic characteristics (Lancefield's system). Hemolytic reactions are based on hemolysins that are produced by streptococci while growing on blood-enriched agar. Blood agar is usually made with defibrinated sheep blood (5.0%), sodium chloride (0.5%) to minimize spontaneous hemolysis, and nutrient agar. Three patterns of hemolysis can occur on blood agar:

1. **β-hemolysis:** Complete hemolysis, giving a clear zone with a clean edge around the colony. (See color plate VIII.5.)

2. **α-hemolysis:** Incomplete hemolysis, producing methemoglobin and a green, cloudy zone around the colony. (See color plate VIII.4.)
3. **γ-hemolysis:** No hemolysis, and no change in the blood agar around the colony.

Streptococci that are α-hemolytic and γ-hemolytic are usually normal microbiota, whereas β-hemolytic streptococci are frequently pathogens.

The streptococci can be antigenically classified into Lancefield groups A through O by antigens in their cell walls. Over 90% of streptococcal infections are caused by β-hemolytic group A streptococci. These bacteria are assigned to the species *S. pyogenes. S. pyogenes* is sensitive to the antibiotic Bacitracin; other streptococci are resistant to Bacitracin (see color plate VII.5). Optochin sensitivity is used to distinguish pathogenic *S. pneumoniae* from other α-hemolytic streptococci.

Sterile blood is added to sterile trypticase soy agar to prepare blood agar. If the mixture is heated to 80°C after addition of the blood, the blood cells lyse, releasing their contents, and the medium turns chocolate brown. This **chocolate agar** is used to culture fastidious bacteria such as *Haemophilus* and some *Neisseria* spp. that require intracellular molecules such as NAD$^+$.

We will examine other bacteria found in the throat in the third part of this exercise.

Materials
Throat Culture
Petri plate containing blood agar

Sterile cotton swab

Gram staining reagents

Hydrogen peroxide, 3% (second period)

Toothpicks

Streptococcus
Petri plate containing blood agar

Sterile cotton swabs (2)

Forceps and alcohol

Optochin disk

Bacitracin disk

10% bile salts (second period)

Tubes containing 2 ml nutrient broth (2) (second period)

Selected Characteristics of Other Bacteria of the Respiratory Tract

Petri plate containing tellurite agar

Petri plate containing brain-heart infusion agar

Gram staining reagents

Methylene blue (simple stain)

Acid-fast staining reagents

CO_2 jar

Oxidase reagent (second period)

Cultures
Streptococcus

Streptococcus pyogenes

Streptococcus pneumoniae

Selected Characteristics of Other Bacteria of the Respiratory Tract

Corynebacterium diphtheriae (avirulent)

Mycobacterium smegmatis on Lowenstein–Jensen medium

Bordetella bronchiseptica on Bordet–Gengou medium

Neisseria sicca on chocolate agar

Techniques Required

Simple staining, Exercise 3

Gram staining, Exercise 5

Acid-fast staining, Exercise 6

Plate streaking, Exercise 11

Catalase test, Exercise 17

Oxidase test, Exercise 17

Kirby–Bauer technique, Exercise 25

Procedure
Throat Culture

 Work only with swabs collected from your own throat.

1. Swab your throat with a sterile cotton swab. The area to be swabbed is between the "golden arches" (glossopalatine arches), as shown in Figure 47.1. Do not hit the tongue.

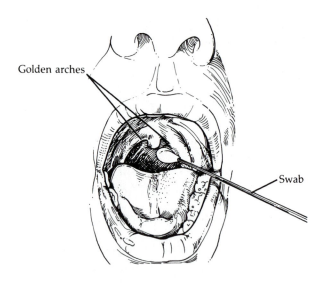

Figure 47.1
Swab your throat with a sterile swab.

2. After obtaining an inoculum from the throat, swab one-half of a blood agar plate. Streak the remainder of the plate with a sterile loop.

 Discard the swab in disinfectant.

3. Incubate the plate, inverted, at 35°C for 24 hours. Observe the plate for hemolysis. Transfer some colonies to a slide with a toothpick, and perform a catalase test. Discard the toothpick in the biohazard container. Why can't the catalase test be done on blood agar? _____
_____
_____

Streptococcus

1. Inoculate each half of a blood agar plate, one side with *S. pyogenes* and the other half with *S. pneumoniae*. Use swabs to obtain confluent growth.

 Be careful—these bacteria are pathogens.

2. Dip forceps in alcohol and burn off the alcohol.

 Keep the forceps tip pointed down while burning. Keep the beaker of alcohol away from the flame.

Using the forceps, place and press a Bacitracin disk and an Optochin disk on each half. Space the disks so that zones of inhibition may be observed.

3. Incubate the plates, inverted, at 35°C for 24 hours. Observe for hemolysis and inhibition of growth by Bacitracin and Optochin.

4. Prepare Gram stains from smears of each organism. Do the organisms differ microscopically? _____

_____

5. Using a sterile loop, prepare a suspension of each organism in a tube of nutrient broth.

6. Add a few drops of 10% bile salts to each tube. Observe the tubes after 15 minutes for lysis of the cells. Which culture is "bile soluble"?_____

Selected Characteristics of Other Bacteria of the Respiratory Tract

1. Perform a Gram stain on smears of each organism, and an acid-fast stain on *Mycobacterium* and one other organism.

2. Record the appearance of *Mycobacterium* on Lowenstein–Jensen medium.

3. Prepare a simple stain of *Bordetella*. Do the cells stain darker in one region?_____

4. Inoculate *Corynebacterium* onto tellurite agar and incubate the plate, inverted, at 35°C for 24 hours. Record any pigment production. Do a simple stain of *Corynebacterium*. Describe the shape of the bacteria.

5. Inoculate brain-heart infusion agar with *Neisseria*. Incubate the plate, inverted, in a CO_2 jar at 35°C for 24 hours. Prepare a CO_2 jar (Figure 47.2) by placing the plates to be incubated with a candle into a large jar. Light the candle and screw the lid on the jar. The candle will extinguish when the CO_2 concentration in the jar reaches 5% to 10%.

6. Perform an oxidase test on *Neisseria*. To test for the production of cytochrome oxidase, follow step a or step b.

   a. Flood an agar culture with oxidase reagent (1% tetramethyl-*p*-phenylene-diamine). Observe for a change in the color of the *colonies*, not the agar. The presence of oxidase is indicated by pink to black colonies.

   b. Moisten an oxidase disk and place it *on* colonies on a plate. Incubate the plate for 15 minutes at 35°C. The disk and colonies turn black in oxidase-positive cultures. (See color plate III.11.)

Figure 47.2
A CO_2 or candle jar. Light the candle and screw the lid onto the jar. Do not let the flame touch the Petri plates. CO_2 jars are used to culture microaerophiles.

exercise 47

Bacteria of the Respiratory Tract

Name _____

Date _____

Lab Section _____

Purpose _____
_____
_____

Data

Throat Culture

	Colony			
	1	2	3	4
Appearance of colonies on blood agar				
Hemolysis				
Catalase reaction				

Streptococcus

	Organism	
	S. pyogenes	*S. pneumoniae*
Gram stain Gram reaction		
Morphology		
Arrangement		
Blood agar plate Appearance of colonies		
Hemolysis		

	Organism	
	S. pyogenes	*S. pneumoniae*
Inhibition by Optochin		
Bacitracin		
Bile solubility		

Other Bacteria

Characteristics	*Mycobacterium smegmatis*	*Bordetella bronchiseptica*	*Cornyebacterium diphtheriae*	*Neisseria sicca*
Gram stain Gram reaction				
Morphology				
Arrangement				
Acid-fast reaction				
Appearance of cells in simple stain	na			na
Oxidase reaction	na	na	na	
Appearance of colonies on Lowenstein–Jensen medium		na	na	na
Bordet–Gengou medium	na		na	na
Tellurite agar	na	na		na
Brain-heart infusion agar	na	na	na	

na = not applicable

Questions

1. Is blood agar selective or differential? _____ Briefly explain. _____

   _____

   Is chocolate agar selective or differential? _____ Briefly explain. _____

   _____

2. Why does addition of a Bacitracin disk to a blood agar plate make a rapid identification technique? _____

   _____

   _____

3. Is the Gram stain of significant importance in the identification of the organisms studied in this exercise? _____

   Explain. _____

   _____

4. You have isolated a gram-positive coccus from a throat culture that you cannot identify as staphylococci or streptococci. A test for one enzyme can be used to distinguish *quickly* between these bacteria. What is the

   enzyme? _____

5. Name one disease caused by each genus used in this exercise. Identify the species that causes that disease.

   _____

   _____

   _____

   _____

   _____

Critical Thinking

1. Design a rapid identification technique for *S. pneumoniae*.

2. Assume that you isolated non-acid-fast, gram-positive, catalase-positive rods from your throat. Does this represent a disease state? Briefly explain.

Learning with Technology

BID Fill in the following table for the organism described in Set 1, Unknown #6. Use the keys in Appendix H and *Bergey's Manual* to identify the bacterial species.

	Results
Appearance on chocolate agar	
Appearance on blood agar	
Gram reaction and morphology	
Glucose fermentation	
Maltose fermentation	
Nitrate reduction	
Sucrose fermentation	
Lactose fermentation	

The bacterium is _____

exercise 48

Bacteria of the Mouth

Objectives

After completing this exercise, you should be able to:

1. List characteristics of streptococci found in the mouth.
2. Describe the formation of dental caries.
3. Explain the relationship between sucrose and caries.

Background

The mouth contains millions of bacteria in each milliliter of saliva. Some of these bacteria are transient microbiota carried on food. Some species of *Streptococcus* are part of the normal microbiota of the mouth. An identification scheme for streptococci found in the mouth is shown in Appendix H, Figure H.4.

Streptococcus mutans, S. salivarius, and *S. sanguis* produce sticky polysaccharides specifically from sucrose. Bacterial exoenzymes hydrolyze sucrose into its component monosaccharides, glucose and fructose. The energy released in the hydrolysis is used by *S. mutans* and *S. sanguis* for polymerization of the glucose to form **dextran.** Fructose released from the sucrose is fermented to produce lactic acid. Enzymes of *S. salivarius* have similar specificity for sucrose, but the fructose is polymerized into **levan,** and the liberated glucose is fermented.

The dextran capsule enables the bacteria to adhere to surfaces in the mouth. *S. salivarius* colonizes the surface of the tongue, and *S. sanguis* and *S. mutans,* the teeth. Masses of bacterial cells, dextran, and debris adhering to the teeth constitute dental **plaque.** Production of lactic acid by bacteria in the plaque initiates dental caries by eroding tooth enamel. Streptococci and other bacteria, such as *Lactobacillus* species, are able to grow on the exposed dentin and tooth pulp.

Other carbohydrates, such as glucose or starch, may be fermented by bacteria, but they are not converted to dextran and hence do not promote plaque formation. "Sugarless" candies contain mannitol or sorbitol, which cannot be converted to dextran, although they may be fermented by such normal microbiota as *S. mutans.* Acid production increases the extent of dental caries.

The number of *S. mutans* present in stimulated saliva has been correlated with the potential for the formation of caries. In this exercise, we will observe and count *S. mutans* and other polysaccharide-producing streptococci in a stimulated saliva sample. The media used contain sucrose to promote capsule formation, and Mitis-Salivarius-Bacitracin (MSB) agar

inhibits the growth of most oral bacteria, except *S. mutans.*

Materials

Rodac plate containing MSB agar

Petri plate containing sucrose blood agar

Chewing gum

Tongue blade

50 ml beaker

1 ml pipettes (3)

Propipette or pipette bulb

99 ml saline dilution blanks (2)

Spreading rod

Beaker containing alcohol

CO_2 jar

3% hydrogen peroxide (second period)

Techniques Required

Catalase test, Exercise 17

Pipetting, Appendix A

Serial dilution technique, Appendix B

Dissecting microscopy, Appendix E

Procedure*

 Work only with your own saliva, and discard all contaminated materials in the To Be Autoclaved area.

Getting Ready

1. Label an MSB plate and a sucrose blood agar plate.
2. Chew a piece of chewing gum until the flavor is gone. Do not swallow your saliva.

*Adapted from C. Hoover. "Indigenous Flora of Saliva and Supragingival Plaque." Unpublished exercise for dental students. San Francisco: University of California, Department of Stomatology, 1991.

MSB Agar

1. Rotate a sterile tongue blade in your mouth ten times so both sides of the blade are thoroughly inoculated with saliva. Remove the tongue blade through closed lips to remove excess saliva. Don't swallow yet.
2. Press one side of the tongue blade onto one-half of the surface of the MSB agar. Then press the other side of the blade onto the other side of the agar, as shown in Figure 48.1.
3. Incubate the plate, inverted, in a CO_2 jar for 72 to 96 hours at 35°C.
4. Prepare a CO_2 jar by placing the plates to be incubated with a candle into a large jar. Light the candle and screw the lid on the jar. The candle will extinguish when the CO_2 concentration in the jar reaches 5% to 10%.
5. Examine the MSB plate with a dissecting microscope (Appendix E) for the presence of *S. mutans*. *S. mutans* colonies are light blue to black, raised, and rough; their surface resembles etched glass or "burnt" sugar. Other bacteria occasionally grow on MSB; therefore, careful observation of colony morphology is necessary to accurately estimate the number of *S. mutans* present.

Sucrose Blood Agar

1. Collect your saliva in a 50 ml beaker. Don't be embarrassed; everyone makes about 2 liters of saliva over the course of a day! Collect approximately 2 to 3 ml. Discard gum in the disinfectant.
2. Label two dilution blanks "1" and "2."
3. Pipette 1 ml of saliva into dilution blank 1. Cap the bottle and shake it. This will give a dilution of 1: _____ . Using another pipette, transfer 1 ml of dilution 1 into dilution blank 2 and invert it several times to mix. This will give a total dilution of 1: _____ .
4. With another pipette, transfer 0.1 ml from dilution 2 to the surface of a sucrose blood agar plate.
5. Disinfect a spreading rod by dipping it in alcohol, quickly igniting the alcohol in a Bunsen burner flame, and letting the alcohol burn off (see Figure 27.3).

> ⚠ **After spreading, dip the spreading rod in alcohol and burn off the alcohol. Keep the flame away from the beaker of alcohol.**

6. Using the cooled spreading rod, spread the fluid over the entire surface of the agar.
7. Disinfect the spreading rod and return it.
8. Incubate the plate, inverted, in a CO_2 jar for 72 to 96 hours at 35°C.

Figure 48.1
To inoculate the Rodac plate, press one side of the tongue blade on the agar. Turn the blade over, and press the other side on the remaining half of the agar.

9. Examine the sucrose blood agar plate for production of large raised mucoid (gum-drop) colonies; these are likely to be *S. salivarius*. Look carefully for glistening colonies surrounded by an indentation of the agar surface that cannot be moved without tearing the agar; these are likely to be *S. sanguis*. *S. mutans* colonies, if present, may be recognized by the presence of a drop of liquid polysaccharide on top of or surrounding the colony. Count the number of each colony type. Record the number of colony-forming units (cfu) per milliliter of saliva.

$$cfu/ml = \frac{Number\ of\ colonies}{Amount\ plated \times dilution}$$

Determine whether catalase is produced by suspected streptococci. The catalase test cannot be done directly on agar containing blood. Why not?

_____

_____

exercise 48

Bacteria of the Mouth

Name _____

Date _____

Lab Section _____

Purpose _____

_____

_____

Data

MSB Agar

Species	Number	Description

Sucrose Blood Agar

	Colony Morphology		
	Suspected *S. salivarius*	Suspected *S. sanguis*	Suspected *S. mutans*
Appearance of colonies			
Number of colonies			
Dilution			
cfu/ml			
Catalase production			

Conclusions

Compare your data with those of your classmates, and draw a conclusion about your oral microbiota. _____

_____

_____

Questions

1. List three characteristics of streptococci found in the mouth. _____

   _____

   _____

2. Studies have shown that both sucrose and bacteria are necessary for tooth decay. Why should this be true? ____

   _____

   _____

   _____

3. Is MSB agar selective or differential? _____ Explain. _____

   _____

4. Is sucrose blood agar selective or differential? _____ Explain. _____

   _____

5. Even if a large number of *S. mutans* is in your saliva, how might you avoid tooth decay? _____

   _____

   _____

   _____

Critical Thinking

1. *S. salivarius* forms large mucoid colonies on sucrose blood agar. Would you expect the same type of colonies on glucose blood agar? Briefly explain your answer.

2. Some dentists determine a patient's susceptibility to dental caries by measuring the pH of the saliva. What is the rationale for using this technique?

Learning with Technology

BID Fill in the following table for the organism described in Set 1, Unknown #25. Use the keys in Appendix H to identify the bacterial species.

	Results
Gram reaction and morphology	
Appearance on blood agar	
OF-glucose reaction	
Catalase reaction	
Insulin fermentation	
Mannitol fermentation	
Acetoin production (V–P test)	

The bacterium is _____

exercise 49

Bacteria of the Gastrointestinal Tract

Objectives

After completing this exercise, you should be able to:

1. List the bacteria commonly found in the gastrointestinal tract.
2. Define the following terms: coliform, enteric, and enterococci.
3. Interpret results from TSI slants and SF broth.

Background

The stomach and small intestine have relatively few microorganisms, as a result of the hydrochloric acid produced by the stomach and the rapid movement of food through the small intestine. But microbial populations in the large intestine are enormous, exceeding 10^{11} bacteria per gram of feces. Most of these intestinal organisms are commensals, and some are in mutualistic relationships with their human hosts. Some intestinal bacteria synthesize useful vitamins, such as folic acid and vitamin K. The normal intestinal microbiota prevent colonization of pathogenic species by producing antimicrobial substances and competition.

The population of the large intestine consists primarily of anaerobes of the genera *Bacteriodes, Bifidobacterium, Enterococcus,* and *Lactobacillus,* and such facultative anaerobes as *Escherichia, Enterobacter, Citrobacter,* and *Proteus.*

Most diseases of the gastrointestinal system result from the ingestion of food or water that contains pathogenic microorganisms. Good sanitation practices, modern methods of sewage treatment, and the disinfection of drinking water help break the fecal–oral cycle of disease. A number of tests have been developed to identify facultative anaerobes associated with fecal contamination. Gram-negative, facultatively anaerobic rods are a large and diverse group of bacteria that includes the **enteric family** (Enterobacteriaceae).

Media have been developed to differentiate between lactose-fermenting enterics and nonlactose-fermenting enterics. The lactose fermenters are called **coliforms** and are generally not pathogenic. The nonlactose-fermenting group includes such pathogens as *Salmonella* and *Shigella.* One of the most common media is MacConkey agar. **MacConkey agar** is selective in that bile salts are inhibitory to gram-positive organisms;

thus, they allow the medium to selectively culture gram-negative organisms. MacConkey agar is differential in that lactose-fermenting organisms (coliforms) give red, opaque colonies and nonlactose fermenters produce colorless, translucent colonies.

After isolation on MacConkey agar, differential screening media such as **triple sugar iron (TSI)** agar can be used to further categorize organisms. TSI contains the following:

0.1% glucose

1.0% lactose

1.0% sucrose

0.02% ferrous sulfate

Phenol red

Nutrient agar

As shown in Figure 49.1, if the organism ferments only glucose, the phenol red indicator will turn yellow in a few hours. The bacteria quickly exhaust the limited supply of glucose and start oxidizing amino acids for energy, giving off ammonia as an end-product. Oxidation of amino acids increases the pH, and the indicator in the slanted portion of the tube will turn back to red. The butt will remain yellow. If the organism in the TSI slant ferments lactose and/or sucrose, the butt and slant will turn yellow and remain yellow for days due to the increased level of acid production. Gas production by an organism can be ascertained by the appearance of bubbles *in* the agar. TSI can also be used to indicate whether hydrogen sulfide (H_2S) has been produced due to desulfurization of sulfur-containing amino acids. H_2S reacts with ferrous sulfate in the medium, producing ferrous sulfide, a black precipitate. The key in Appendix H, Figure H.5, shows how enteric bacteria can be identified using MacConkey agar, TSI, and additional tests.

The presence of enterococci can be used to indicate fecal contamination. Enterococci are fairly specific for their mammalian hosts. *Enterococcus faecalis* (formerly called "Streptococcus faecalis") is found in humans, *S. bovis* in cattle, and *S. equinus* in horses. **SF (*Streptococcus faecalis*) broth** can be used to detect the presence of enterococci. SF broth contains sodium azide to inhibit the growth of gram-negative

(a) No sugar fermentation

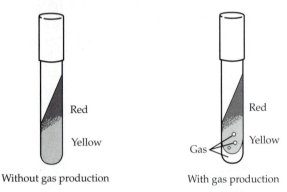

(b) Glucose fermented; lactose and sucrose not fermented

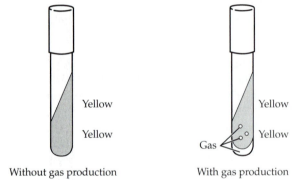

(c) Glucose and lactose and/or sucrose fermented

(d) H$_2$S production can occur in addition to **(a)**, **(b)**, and **(c)**.

Figure 49.1
Reactions in triple sugar iron (TSI) agar after incubating for 24 hours.

bacteria. Enterococci will grow and ferment the glucose in the broth. Growth and a change in the bromcresol purple indicator from purple (pH 6.6) to yellow (pH 5.2) may indicate the presence of enterococci.

Tomato juice agar is an enriched medium used to encourage the growth of lactobacilli. *Lactobacillus* does not grow well on ordinary media, but its growth is enhanced by the addition of tomato juice and peptonized milk.

Materials

Petri plate containing MacConkey agar

Petri plate containing tomato juice agar

Tube containing SF broth

Sterile water

Brewer anaerobic jar and GasPak®

Tube containing sterile cotton swab

10 ml pipette

Fecal sample (animal or human feces; swab feces or the anus after a bowel movement with a sterile swab, and bring the contaminated swab to class in a sterile tube)

Second period:
Gram stain reagents

TSI slant

3% hydrogen peroxide

Techniques Required

Gram staining, Exercise 5

Plate streaking, Exercise 11

Fermentation tests, Exercise 14

Catalase test, Exercise 17

Anaerobic culture techniques, Exercise 19

Procedure

1. If the feces are hard, break them apart in a sterile Petri plate and emulsify them in sterile water. Swab a small area on the MacConkey agar with an inoculated swab, then streak for isolation with a loop (Figure 46.1b).
2. Incubate the plate, inverted, at 35°C for 24 to 48 hours.
3. Using the fecal swab, inoculate tomato juice agar. Incubate the plate, inverted, anaerobically at 35°C for 24 to 48 hours.

4. Inoculate SF broth by placing the fecal swab in the broth. Incubate the SF broth at 35°C for 24 to 48 hours.

> **Discard the test tube, tube of water, and other contaminated materials in the To Be Autoclaved area.**

5. Observe your MacConkey plate and those of other students. Record the appearance and colors of the colonies. (See color plate XI.6.) Prepare a Gram stain from an isolated colony.

6. Select one isolated colony using an inoculating needle and inoculate the TSI slant. *Stab* into the butt of the agar, then streak the surface of the slant. Incubate the tube at 35°C for 24 hours. Observe and record results from the TSI slant inoculated. (See color plate III.12.) Observe other students' TSI slants.

7. Observe the tomato juice agar for the presence of *Lactobacillus*. Perform a Gram stain and a catalase test on isolated colonies. Is *Lactobacillus* catalase-positive or catalase-negative? _____

8. Observe the SF broth for the presence of entero-cocci. Perform a Gram stain.

exercise 49

Bacteria of the Gastrointestinal Tract

Name _____

Date _____

Lab Section _____

Purpose _____

_____

_____

Data

MacConkey Agar

Source of inoculum: _____

	Colony			
	1	2	3	4
Appearance of colonies				
Lactose fermentation?				

TSI

Colony picked: _____

Appearance: Slant _____

Butt _____

Gas? _____

H$_2$S _____

What reactions occurred in the TSI? _____

_____

_____

_____

_____

Tomato Juice Agar

Description of colonies: _____

_____

Gram reaction of colony picked: _____

Catalase test: _____

Cultures

Tube containing *Escherichia coli, Proteus vulgaris,* and *Pseudomonas aeruginosa*

G.C. Smears

G.C. smears

One unknown smear # _____

Techniques Required

Gram staining, Exercise 5

Plate streaking, Exercise 11

Fermentation tests, Exercise 14

Urea hydrolysis, Exercise 15

Oxidase test, Exercise 17

MacConkey agar, Exercise 49

Pipetting, Appendix A

Serial dilution technique, Appendix B

Procedure

Urine Culture

 Work only with urine collected from your own body.

1. Collect a "clean-catch" urine specimen, as described by your instructor, using the sterile jars available in the lab. Refrigerate the specimen until you are ready to perform step 2. Why?_____

    _____

2. Add 0.1 ml of urine to one dilution blank, and pipette up and down three times to mix. With a new pipette, transfer 0.1 ml to the second tube and mix. Transfer 0.1 ml to the surface of the blood agar plate. This is a 1: _____ dilution.

3. Dip the spreading rod in alcohol, and flame off the excess alcohol to disinfect. Spread the 0.1 ml of diluted urine on the blood agar with the spreading rod (see Figure 27.3).

    ⚠️ **After spreading, dip the spreading rod in alcohol and ignite it. Keep the flame away from the beaker of alcohol.**

4. Streak a loopful of *undiluted* urine on MacConkey agar.

 Discard the jar and pipettes in the appropriate biohazard container.

5. Incubate both plates, inverted, at 35°C for 24 to 48 hours.

6. Count the colonies on the blood agar plate, and determine the number of colony-forming units (cfu) per milliliter of urine:

    $$\text{cfu/ml} = \frac{\text{Number of colonies}}{\text{Amount plated} \times \text{dilution}}$$

7. Examine the MacConkey plate for the presence of possible coliforms. Consult your instructor if you are alarmed by your results.

Cystitis

1. Inoculate a MacConkey plate and Pseudomonas agar P plate with the mixed culture of bacteria.

2. Incubate the plates, inverted, at 35°C for 24 to 48 hours. (See color plates V.3, V.4, and XI.6.)

3. Examine the MacConkey plate for lactose-fermenting colonies. Which of the three organisms ferments lactose? _____
    Look for swarming. Which organism is actively motile on solid media?_____
    Are any small, nonlactose-fermenting colonies present? _____

4. Prepare a Gram stain from each different colony type.

5. Examine the Pseudomonas agar P plate. Can you identify *Pseudomonas*? _____

6. Perform an oxidase test on each different organism.

7. Inoculate four tubes of OF-glucose medium: two with *Proteus* and two with *Pseudomonas*. Plug one tube of each organism with mineral oil. Inoculate a urea agar slant with each organism. Incubate the tubes at 35°C for 24 to 48 hours.

8. Record the results of the OF-glucose and urease production tests. Why isn't it necessary to perform these two biochemical tests on *E. coli*? _____

    _____

G.C. Smears

1. Examine the G.C. smears provided.

2. Determine whether *N. gonorrhoeae* could be present in your unknown G.C. smear.

exercise 50

Bacteria of the Urogenital Tract

Name _____

Date _____

Lab Section _____

Purpose _____

_____

Data

Urine Culture

Blood agar plate:

  Number of colonies: _____

  Total count (cfu/ml): _____

  Hemolysis: _____

MacConkey agar plate:

  Possible coliforms present: _____

  If so, describe the colonies: _____

  Noncoliforms: _____

Cystitis

	Organism		
	E. coli	*P. vulgaris*	*P. aeruginosa*
MacConkey agar			
Appearance of colonies	_____	_____	_____
Lactose fermentation	_____	_____	_____
Swarming	_____	_____	_____
Gram stain	_____	_____	_____
Pseudomonas agar P			
Appearance of colonies	_____	_____	_____
Pigmentation	_____	_____	_____
Oxidase reaction	_____	_____	_____
OF-glucose (fermentation or oxidation)	_____	_____	_____
Urease production	_____	_____	_____

G.C. Smears

Diagram the appearance of a known G.C.-positive smear.

Diagram the appearance of a known G.C.-negative smear.

Unknown # _____
Diagram the appearance of your unknown smear.

_____ ×

_____ ×

_____ ×

Could *N. gonorrhoeae* be present? _____

Questions

1. Why is MacConkey agar inoculated with a urine specimen? _____
   _____
   _____

2. With the results obtained on the blood agar and MacConkey agar plates, is a urinary tract disease possible? _____
   _____

3. The enterics and pseudomonads look alike microscopically. How can you easily distinguish between these two
   groups of bacteria? _____
   _____
   _____

4. Why are females more prone to urinary tract infections than males? _____
   _____
   _____

Critical Thinking

1. Differentiate between the pigment of *P. aeruginosa* on Pseudomonas agar P and the "pigment" of *E. coli* on MacConkey agar without referring to the colors.

2.   What role does antibiotic treatment have in yeast infections of the urinary tract?

Learning with Technology

BID Fill in the following table for the organism described in Set 1, Unknown #30. Use the keys in Appendix H and *Bergey's Manual* to identify the bacterial species.

	Results
Gram reaction and morphology	
Appearance on MacConkey agar	
Oxidase reaction	
OF-glucose reaction	
Indole production	
Methyl red test	
Acetoin production (V–P test)	
Citrate utilization	
β-galactosidase	
Lactose fermentation	
H$_2$S production	

The bacterium is _____

exercise 51

Identification of an Unknown from a Clinical Sample

Unknown #3*

So simple did it seem to me
The day t'was first presented,
A Sherlock Holmes of lab technique
No effort be resented.

Now six Gram stains, three cultures later
The first step barely taken
Gram vari rods, both red and blue
My confidence was shaken.

Straight rods 'tis true, yet red and blue
In ones and twos and chain
Somes spores would show the truth of it
But stains were all in vain.

Alas! No spores of vivid green
Would show in fields of red.
Ah ha! Too soon! Spores take some time
Flashed through my seeking head.

With rod in hand, though young it be,
It surely must, said I,
Reveal some trait, another test
Than just a non-cocci.

So while I wait don't waste the time
The sugars show some trait.
So off to these, ferment or not
I can't just sit and wait.

Glucose a plus, Sucrose the same
No gas, and lactose zilch.
The spores again with no green spots
It's on to litmus milk.

So mannitol and nitrate too
I'll run, for these should show,
Though spores resist at now day six,
Some little cells should grow.

Day seven's gone and spores still hide
From vision sharply tooled.
But agents stained from tests complete
Have many others ruled.

Divorce now near, by child disowned
On job, the thought "to fire"
I'll take my stand and flunk or not
B. subtilis, or tests are liars.

MARY DAVIS

Objectives

After completing this exercise, you should be able to:

1. Separate and identify two bacteria from a sample.
2. Determine the sensitivity of the unknown bacteria to chemotherapeutic agents.

Background

Clinical samples often contain many microorganisms, and a pathogen must be separated from resident normal microbiota. Differential and selective media are used to facilitate isolation of the pathogen, after which it must be identified. Also, antibiotic sensitivity tests are performed to aid the physician in prescribing treatment. Since effective treatment depends on identification of the pathogen, a clinical laboratory must provide results as quickly and accurately as possible. Microbiology laboratory results are usually available within 48 hours.

*M. E. Davis. *ASM News* 42(3):164, 1976. Reprinted with the permission of the American Society for Microbiology.

In this exercise, you will be provided with a simulated clinical sample containing two bacteria. The procedure will differ from that used in a clinical laboratory in that you will be asked to isolate and identify *both* organisms in the sample. In a clinical laboratory, selective and enrichment techniques are used to identify the pathogen only. After you have separated the two bacteria in pure culture, prepare stock and working cultures. To obtain pure cultures, inoculate a trypticase soy agar plate *and* a differential medium. Select the differential medium using the type of sample as a clue. For instance, a fecal sample should probably be inoculated onto _____ _____ agar.

Since differential media should give the information needed to isolate the organisms, why is a trypticase soy agar inoculated also? _____

Materials

Petri plates containing trypticase soy agar

All stains, reagents, and media previously used

Trypticase soy agar slants (second period)

Mueller–Hinton agar (third period)

Antimicrobial disks (third period)

Culture

Unknown sample # _____

Techniques Required

Gram staining, Exercise 5

Plate streaking, Exercise 11

Differential and selective media, Exercise 12

Kirby–Bauer technique, Exercise 25

Use of *Bergey's Manual,* Exercise 33

Procedure

Getting Started

1. Review general information regarding identification of an unknown. Study the identification schemes in Appendix H, Figures H.1 through H.5.
2. You may be able to prepare a Gram stain directly from the unknown if it does not contain a high concentration of organic matter to interfere with the stain results.
3. Streak the unknown onto a trypticase soy agar plate and an appropriate differential medium. *Do not be wasteful.* Inoculate only the necessary media. Incubate the plates at 35°C for 24 to 48 hours.

Second Period

1. Examine the plates, and select colonies that differ from each other in appearance and that are separated for easy isolation. (See color plate XII.)
2. Streak each organism for isolation onto half of a trypticase soy agar plate. Incubate the plate at 35°C for 24 to 48 hours. This plate can be your first working culture. Prepare a stock culture of each organism on separate trypticase soy agar slants (one culture on each slant).

Third Period

1. Prepare a Gram stain of each organism. What should you do if you do not think you have a pure culture? _____
   _____
   _____

2. After determining the staining and morphologic characteristics of the two organisms, determine which biochemical tests you will need. Plan your work carefully. The same tests need not be performed on both bacteria.
3. Record which tests were performed and the results.
4. Select six antibiotics from Table 25.1 to test against your unknowns. Use the Mueller–Hinton agar and antimicrobial disks to determine whether your unknowns are resistant to any antimicrobial agents. (See color plate VIII.1.)

exercise 51

Identification of an Unknown from a Clinical Sample

Name _____

Date _____

Lab Section _____

Purpose _____

_____

_____

Data

Unknown # _____

Source: _____

Preliminary Gram stain results: _____

_____

	Bacterium	
	A	B
Appearance of colonies on trypticase soy agar		
Gram reaction		
Cell morphology and arrangement		

Make a flowchart of your procedure. Remember, the *same* tests may *not* be required for each bacterium. The following will help you get started:

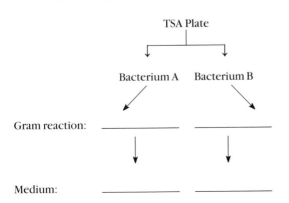

	Zone of Inhibition (mm)	
	Bacterium A	Bacterium B
Antibiotic sensitivity to		
_____	_____	_____
_____	_____	_____
_____	_____	_____
_____	_____	_____
_____	_____	_____
_____	_____	_____

Conclusions

1. What two bacterial species were in unknown # _____ ? _____

    _____

2. Are either of your unknowns resistant to any antibiotics? _____

    If so, which ones? _____

    _____

    _____

Questions

1. Write your rationale for arriving at the conclusion.

2. Why would a clinical microbiology laboratory perform antimicrobial susceptibility tests as well as identify the

    unknown? _____

Microbiology and the Environment

EXERCISES

Microorganisms are omnipresent in our environment; however, the presence of some microorganisms in certain places, such as food or water, may be undesirable. In Exercises 52 and 53, we will use the presence of nonpathogenic bacteria to indicate water pollution. The presence and number of bacteria in foods can indicate that foods were contaminated during processing; these bacteria may result in food spoilage (Exercise 54).

Harmful microorganisms are only a very small fraction of the total microbial population. The activities of most microbes are in fact beneficial. In Exercise 55, we will use selected microorganisms to produce desired flavors in foods and to preserve foods.

The activities of soil bacteria are essential to the maintenance of life on Earth (see the illustration on the next page). Nitrogen fixation (Exercise 56) was first described in 1893 by Sergei Winogradsky, who entered into the study of biogeochemical cycles because he was *impressed by the incomparable glitter of Pasteur's discoveries. I started to investigate the great problem of fixation of atmospheric nitrogen. I succeeded without too much difficulty in isolating an anaerobic rod called* Clostridium *that would perform this function.**

Winogradsky named the organism *Clostridium pasteurianum* in honor of Pasteur. Some microbes are able to degrade dangerous pollutants in the environment (Exercise 57).

*Quoted in H. A. Lechevalier and M. Solotorovsky. *Three Centuries of Microbiology.* New York: Dover Publications, 1974, p. 263.

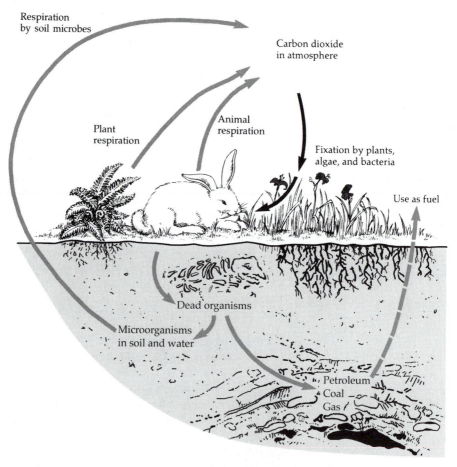

The carbon cycle.

exercise 52

Microbes in Water: Multiple-Tube Technique

Objectives

After completing this exercise, you should be able to:

1. Define coliform.
2. Provide the rationale for determining the presence of coliforms.
3. List and explain each step in the multiple-tube technique.

Background

Tests that determine the bacteriologic quality of water have been developed to prevent transmission of waterborne diseases of fecal origin. However, it is not practical to look for pathogens in water supplies, because pathogens occur in such small numbers that they might be missed by sampling. Moreover, when pathogens are detected, it is usually too late to prevent occurrence of the disease. Rather, the presence of **indicator organisms** is used to detect fecal contamination of water.

An indicator organism must be present in human feces in large numbers and must be easy to detect. The most frequently used indicator organisms are the coliform bacteria. **Coliforms** are aerobic or facultatively anaerobic, gram-negative, nonendospore-forming, rod-shaped bacteria that ferment lactose with acid and gas formation within 48 hours at 35°C. Coliforms are not usually pathogenic, although they can cause opportunistic infections. Coliforms are not restricted to the human gastrointestinal tract but may be found in other animals and in the soil. Tests that determine the presence of fecal coliforms (of human origin) have been developed.

Established public health standards specify the maximum number of coliforms allowable in each 100 ml of water, depending on the intended use of the water (for example, drinking, water-contact sports, or treated wastewater for irrigation or for discharge into a bay or river).

Coliforms can be detected and enumerated in the **multiple-tube technique** (Figure 52.1). In this method,

Figure 52.1

Analysis of drinking water for coliforms by the multiple-tube technique. *Source: Basic Microbiology with Applications,* 2nd Edition, by T. D. Brock and K. M. Brock, copyright © 1978. Reprinted with permission of Prentice-Hall, Inc., Upper Saddle River, NJ.

Table 52.1

Most Probable Numbers (MPN) Index for Various Combinations of Positive and Negative Results When Three 10 ml Portions, Three 1 ml Portions, and Three 0.1 ml Portions Are Used

Number of tubes giving positive reaction out of			MPN Index per 100 ml	Number of tubes giving positive reaction out of			MPN Index per 100 ml
3 of 10 ml Each	3 of 1 ml Each	3 of 0.1 ml Each		3 of 10 ml Each	3 of 1 ml Each	3 of 0.1 ml Each	
0	0	0	<3	3	0	0	23
0	0	1	3	3	0	1	39
0	1	0	3	3	0	2	64
1	0	0	4	3	1	0	43
1	0	1	7	3	1	1	75
1	1	0	7	3	1	2	120
1	1	1	11	3	2	0	93
1	2	0	11	3	2	1	150
2	0	0	9	3	2	2	210
2	0	1	14	3	3	0	240
2	1	0	15	3	3	1	460
2	1	1	20	3	3	2	1,100
2	2	0	21	3	3	3	≥2,400
2	2	1	28				

Source: Standard Methods for the Examination of Water and Wastewater, 13th ed. New York: American Public Health Association, 1971.

coliforms are detected in three stages. In the **presumptive test,** dilutions from a water sample are added to lactose fermentation tubes. The lactose broth can be made selective for gram-negative bacteria by the addition of lauryl sulfate or brilliant green and bile. Fermentation of lactose to gas is a positive reaction.

Samples from the positive presumptive tube at the highest dilution are examined for coliforms by inoculating a differential medium in the **confirmed test.** A confirmed test can be done on **MUG agar.** Almost all strains of *E. coli* produce the enzyme β-glucuronidase (GUD). If *E. coli* is added to a nutrient medium containing 4-methylumbelliferone glucuronide (MUG), GUD converts MUG to a fluorescent compound that is visible with an ultraviolet lamp (see color plate XI.5).

The number of coliforms is determined by a statistical estimation called the **most probable number (MPN)** method. In the presumptive test, tubes of lactose broth are inoculated with samples of the water being tested. A count of the number of tubes showing acid and gas is then taken, and the figure compared to statistical tables, shown in Table 52.1. The number is

the most probable number of coliforms per 100 ml of water.

Materials

9 ml single-strength lactose fermentation tubes (6)

20 ml 1.5-strength lactose fermentation tubes (3)

Petri plate containing MUG agar (second period)

Sterile 10 ml pipette

Sterile 1 ml pipette

Gram staining reagents

Water sample, 50 ml (Bring your own, from a pond or stream.)

Techniques Required

Gram staining, Exercise 5

Aseptic technique, Exercise 10

Plate streaking, Exercise 11

Fermentation tests, Exercise 14

Pipetting, Appendix A

Procedure

1. Label three single-strength lactose broth tubes "0.1," label another three tubes "1," and label the three 1.5-strength broth tubes "10."
2. Inoculate each 0.1 tube with 0.1 ml of your water sample.
3. Inoculate each 1 tube with 1.0 ml of your water sample.
4. Inoculate each 10 tube with 10 ml of the water sample. Why is 1.5-strength lactose broth used for this step? _____ _____
5. Incubate the tubes for 24 to 48 hours at 35°C.

6. Record the results of your presumptive test (Figure 52.1). Which tube has the highest dilution of the water sample? _____ _____

    Determine the number of coliforms per 100 ml of the original sample using Table 52.1. If a tube has gas, streak the MUG agar with the positive broth. Incubate the plate, inverted, for 24 to 48 hours at 35°C.
7. Examine the plate using an ultraviolet lamp.

> ⚠ **Do not look directly at the ultraviolet light, and do not leave your hand exposed to it.**

Record the results of your confirmed test (Figure 52.1). (See color plate XI.5.) How can you tell whether coliform colonies are present? _____ _____

exercise 52

Microbes in Water: Multiple-Tube Technique

Name _____

Date _____

Lab Section _____

Purpose _____

_____

_____

Data

Water sample source: _____

Presumptive Test

| Inoculum | Number of tubes with: | | | Possible coliforms present? |
	Growth	Acid	Gas	
0.1 ml				
1.0 ml				
10 ml				

MPN: _____

Confirmed Test

Tube: _____ Growth: _____

Appearance of colonies: _____

_____

Are coliforms present? _____

Data from water samples tested by other students:

Sample	MPN	Coliforms present?

Conclusions

1. Were coliforms present in your water sample? _____

2. What is the MPN of your water sample? _____ per _____ ml

Questions

1. Could the water have a high concentration of the pathogenic bacterium *Vibrio cholerae* and give negative results

    in the multiple-tube technique? Briefly explain. _____

    _____

2. Why are coliforms used as indicator organisms if they are not usually pathogens? _____

    _____

3. Why isn't a pH indicator needed in the lactose broth fermentation tubes? _____

    _____

Critical Thinking

1. Why didn't we inoculate MUG agar directly and bypass lactose broth?

2. The following table lists the etiologies of illness associated with drinking and recreational waters in 1994. Which of the water samples would have shown a high coliform count? Which diseases could have been prevented by chlorinating the water? Briefly explain your answers.

Agent	Number of Cases
Campylobacter jejuni	223
Cryptosporodium parvum	403,964
E. coli O157:H7	166
Giardia lamblia	526
Naegleria fowleri	1
Shigella flexneri	68
Shigella sonnei	632

3. Use the data shown below to explain how waterborne diseases can best be eliminated.

exercise 53

Microbes in Water: Membrane Filter Technique

Objectives

After completing this exercise, you should be able to:

1. Explain the principle of the membrane filter technique.
2. Perform a coliform count using the membrane filter technique.

Background

Fecal contamination of water can be determined by the number of coliforms present in a water sample, by the multiple-tube technique (Exercise 52). Coliforms can also be detected by the membrane filter technique.

In the **membrane filter technique,** water is drawn through a thin filter (see Appendix F). Filters with a variety of pore sizes are available. Pores of 0.45 μm are used for filtering out most bacteria. Bacteria are retained on the filter, which is then placed on a pad of suitable nutrient medium. Nutrients that diffuse through the filter can be metabolized by bacteria trapped on the filter. Each bacterium that is trapped on the filter will develop into a colony. Bacterial colonies growing on the medium can be counted. When a selective or differential medium is used, desired colonies will have a distinctive appearance. EMB (eosin methylene blue) agar is frequently used as a selective and differential medium with the membrane filter technique. The composition of EMB agar is shown in Table 53.1. **EMB agar** is selective because the EMB dyes inhibit the growth of gram-positive organisms, allowing the growth of gram-negative bacteria. EMB is differential in that the colonies of lactose-fermenting bacteria have a pink to blue color, black center, or metallic sheen from the eosin and methylene blue (see color plates XI.1, XI.2, XI.3, XI.4, and XI.7).

Materials

47 mm Petri plate containing EMB agar

Sterile membrane filter apparatus

Sterile 0.45 μm filter

Forceps

Alcohol

Sterile pipette or graduated cylinder, as needed

Table 53.1
Chemical Composition of Bacto EMB Agar

Peptone	1.0%
Lactose	0.5%
Sucrose	0.5%
Dipotassium phosphate	0.2%
Agar	1.35%
Eosin Y	0.04%
Methylene blue	0.0065%

Source: Difco Laboratories.

Sterile rinse water

Water sample (Bring your own, from a pond or stream.)

Techniques Required

Pipetting, Appendix A

Membrane filtration, Appendix F

Procedure

1. Set up filtration equipment (see Appendix F). Remove wrappers as each piece is fitted into place. Why shouldn't all the wrappers be removed at once? _____

   a. Attach the filter trap to the vacuum source. What is the purpose of the filter trap? _____

   b. Place the filter holder base (with stopper) on the filtering flask. Attach the flask to the filter trap.

   ⚠ **Disinfect the forceps by burning off the alcohol. Keep the beaker of alcohol away from the flame.**

   c. Using the sterile forceps, place a filter on the filter holder. Why must the filter be centered exactly on the filter holder? _____

    d. Set the funnel on the filter holder and fasten it in place.

2. Filtering

    a. Shake the water sample, and pour or pipette a measured volume into the funnel. Your instructor will help you determine the volume. (*For samples of 10 ml or less, pour 20 ml sterile water into the funnel first.*)

    b. Turn on the vacuum and allow the sample to pass into the filtering flask. Leave the vacuum on.

    c. Pour sterile rinse water into the funnel. (*Use the same volume as the sample.*) Allow the rinse water to go through the filter. Turn the vacuum off.

3. Inoculation (Figure 53.1)

    a. Carefully remove the filter from the filter holder using sterile forceps. Why does the filter have to be "peeled" off? _____

    _____

    b. Carefully place the filter on the culture medium. Do not bend the filter; place one edge down first, then carefully set the remainder down. Place the filter on the plate as it was in the filter holder.

4. Invert the plate and incubate it for 24 hours at 35°C.

5. Examine the plates for the presence of coliforms. On EMB, coliforms will form colored colonies, and some may have a green metallic sheen. (See color plates XI.1, XI.2, XI.3, XI.4, and XI.7.) Count the number of coliform colonies:

Number of coliforms per 100 ml of water =

$$100 \times \frac{\text{Number of coliform colonies}}{\text{Volume of sample filtered}}$$

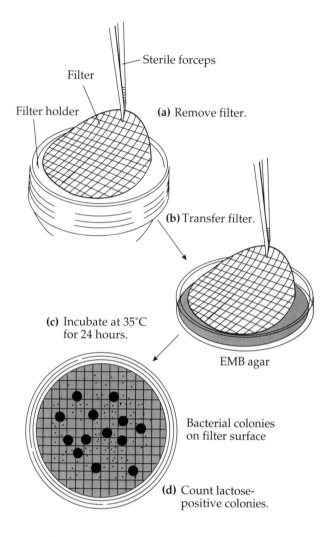

(a) Remove filter.

(b) Transfer filter.

(c) Incubate at 35°C for 24 hours.

EMB agar

Bacterial colonies on filter surface

(d) Count lactose-positive colonies.

Figure 53.1

Inoculation. Using sterile forceps, remove the filter from the filter holder. Place the filter on the culture medium, gradually laying it down from one edge to the other.

exercise 53

Microbes in Water: Membrane Filter Technique

Name _____

Date _____

Lab Section _____

Purpose _____

_____

_____

Data

Sample tested: _____

Describe the general appearance of the colonies on EMB medium: _____

_____

_____

_____

Number of coliform colonies: _____

Number of coliforms/100 ml water: _____
Use the space below for your calculations.

Data from water samples tested by other students:

Sample	Coliforms/100 ml	

Conclusions

1. Which water sample(s) is (are) potable? _____

2. Which water sample(s) is (are) contaminated with fecal material? _____

Questions

1. What basic assumption is made in this technique, if the number of bacteria is determined from the number of

    colonies? _____

    _____

2. Why can filtration be used to sterilize culture media? _____

    Sewage? _____

    Air? _____

3. If you did the multiple-tube technique, list one advantage and one disadvantage of each method of detecting

    coliforms. _____

    _____

    _____

    _____

4. Why is the membrane filter technique useful for a sanitarian working in the field? _____

    _____

    _____

Critical Thinking

1. Can you determine the source of contamination (for example, human, domestic animal, or wild animal) from
    this test?

2. Design an experiment to detect the presence and number of *Acetobacter* in wine. Why would you want to
    perform such a test?

exercise 54

Microbes in Food: Contamination

Objectives

After completing this exercise, you should be able to:

1. Determine the approximate number of bacteria in a food sample using a standard plate count.
2. Provide reasons for monitoring the bacteriologic quality of foods.
3. Explain why the standard plate count is used in food quality control.

Background

Illness and food spoilage can result from microbial growth in foods. The sanitary control of food quality is concerned with testing foods for the presence of pathogens. During processing (grinding, washing, and packaging), food may be contaminated with soil microbes and microbiota from animals, food handlers, and machinery.

Foods are the primary vehicle responsible for the transmission of diseases of the digestive system. For this reason, they are examined for the presence of coliforms because the presence of coliforms usually indicates fecal contamination.

Standard plate counts are routinely performed on food and milk by food-processing companies and public health agencies. The **standard plate count** is used to determine the total number of viable bacteria in a food sample. The presence of large numbers of bacteria is undesirable in most foods because it increases the likelihood that pathogens will be present, and it increases the potential for food spoilage.

In a standard plate count, the number of **colony-forming units (cfu)** is determined. Each colony may arise from a group of cells rather than from one individual cell. The initial sample is diluted through serial dilutions (Appendix B) in order to obtain a small number of colonies on each plate. A known volume of the diluted sample is placed in a sterile Petri plate and melted, and cooled nutrient agar is poured over the inoculum. After incubation, the number of colonies is counted. Plates with between 25 and 250 colonies are suitable for counting. A plate with fewer than 25 colonies is unsuitable for counting because a single contaminant could influence the results. A plate with more than 250 colonies is extremely difficult to count. The microbial population in the original food sample can then be calculated using the following equation:

Colony-forming units/gram or ml of sample =

$$\frac{\text{Number of colonies}}{\text{Amount plated} \times \text{dilution}^*}$$

A limitation of the standard plate count is that only bacteria capable of growing in the culture medium and environmental conditions provided will be counted. A medium that supports the growth of most heterotrophic bacteria is used.

Materials

Melted standard plate count or nutrient agar, cooled to 45°C

Sterile 1 ml pipettes (6)

Sterile Petri plates (8)

Sterile 99 ml dilution blanks (3)

Food samples, diluted 1:10

Techniques Required

Aseptic technique, Exercise 10

Pour plates, Exercise 11

Pipetting, Appendix A

Serial dilution technique, Appendix B

Procedure

First Period. A: Bacteriologic Examination of Milk

1. Obtain a sample of either raw or pasteurized milk that has been diluted 1:10.
2. Using a sterile 1 ml pipette, aseptically transfer 1 ml of the 1:10 milk sample into a 99 ml dilution blank; label the bottle "1:10³" and discard the pipette (Figure 54.1a). Shake the bottle 20 times, with your elbow resting on the table, as shown in Figure 54.2.

*"Dilution" refers to the tube prepared by serial dilutions (Appendix B). For example, if 250 colonies were present on the 1:10⁶ plate, the calculation would be as follows:

$$\text{Colony-forming units/gram} = \frac{250 \text{ colonies}}{0.1 \text{ ml} \times 10^{-5}}$$
$$= 250 \times 10^5 \times 10^1$$
$$= 250,000,000$$
$$= 2.5 \times 10^8$$

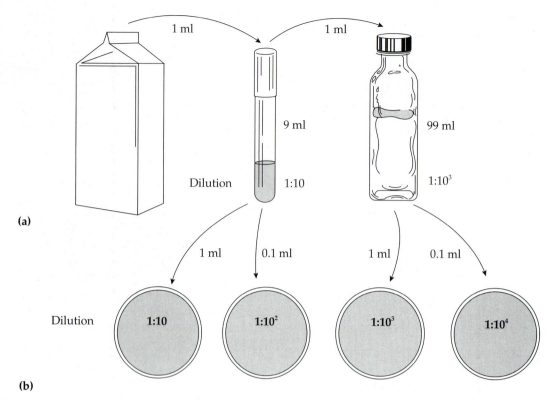

(a)

(b)

Figure 54.1

Standard plate count of milk. (a) Make serial dilutions of a milk sample. (b) Mark four Petri plates with the dilutions.

3. Label the bottoms of four sterile Petri plates with the dilutions: "1:10," "1:10²," "1:10³," and "1:10⁴" (Figure 54.1b).

4. Using a 1 ml pipette, aseptically transfer 0.1 ml of the 1:10³ dilution into the bottom of the 1:10⁴ plate. *Note:* 0.1 ml of the 1:10³ dilution results in a 1:10⁴ dilution of the original sample. Using the same pipette, transfer 1.0 ml of the 1:10³ dilution into the plate labeled 1:10³. Pipette 0.1 ml and 1.0 ml from the 1:10 dilution into the 1:10² and 1:10 plates, respectively, with the same pipette (Figure 54.3a). Why is it important to proceed from the highest to the lowest dilution? _____
_____

5. Check the temperature of the water bath containing the nutrient agar. Test the temperature of the outside of the agar container with your hand. It should be "baby bottle" warm. Why? _____
_____

6. Pour the melted nutrient agar into one of the plates (to about one-third full) (Figure 54.3b). Cover the plate and swirl it gently (Figure 54.3c) to distribute the milk sample evenly through the agar. Continue until all the plates are poured.

7. When each plate has solidified, invert it, and incubate all plates at 35°C for 24 to 48 hours.

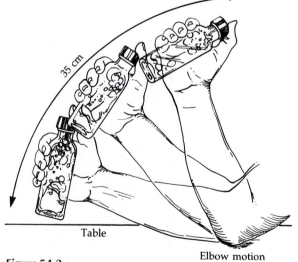

Figure 54.2

Shake the dilution bottle 20 times through a 35 cm arc.

First Period. B: Bacteriologic Examination of Hamburger and Frozen Vegetables

1. Obtain a sample of raw hamburger or frozen vegetables diluted 1:10.

Figure 54.3

Pour plate. **(a)** Pipette inoculum into a Petri plate. **(b)** Add liquefied nutrient agar, and **(c)** mix the agar with the inoculum by gently swirling the plate.

Figure 54.4

Standard plate count on food. **(a)** Prepare serial dilutions of a food sample. **(b)** Label four Petri plates with the dilutions to be plated.

2. Using a sterile 1 ml pipette, aseptically transfer 1 ml of the 1:10 sample into a 99 ml dilution blank; label the bottle "1:10³" and discard the pipette. Shake the bottle 20 times, with your elbow resting on the table, as shown in Figure 54.2. Make a 1:10⁵ dilution using another 99 ml dilution blank (Figure 54.4a). Shake the bottle as before.

3. Label the bottoms of four sterile Petri plates with the dilutions: "1:10³," "1:10⁴," "1:10⁵," and "1:10⁶" (Figure 54.4b).

4. Using a 1 ml pipette, aseptically transfer 0.1 ml of the 1:10⁵ dilution into the 1:10⁶ plate. *Note:* 0.1 ml of a 1:10⁵ dilution results in a 1:10⁶ dilution of the original sample. Using the same pipette, repeat this procedure with the 1:10³ dilution until all the plates have been inoculated (Figure 54.3a). Why can the same pipette be used for each transfer? _____

_____

_____

5. Check the temperature of the water bath containing the nutrient agar. Test the temperature of the outside of the agar container with your hand. It should be "baby bottle" warm. Why? _____

_____

6. Pour the melted nutrient agar into one of the plates (to about one-third full) (Figure 54.3b). Cover the plate and swirl it gently (Figure 54.3c) to distribute the sample through the agar evenly. Continue until all the plates are poured.

7. When each plate has solidified, invert it, and incubate all plates at 35°C for 24 to 48 hours.

Second Period

1. Arrange each plate in order from lowest to highest dilution.

2. Select the plate with 25 to 250 colonies. Record data for plates with fewer than 25 colonies as *too few to count (TFTC)* and those with more than 250 colonies, *too numerous to count (TNTC)*.

3. Count the number of colonies on the plate selected.

4. Calculate the number of bacteria in the original food. For example, if 129 colonies were counted on a 1:10³ dilution:

$$\frac{129 \text{ colonies}}{1 \text{ ml} \times 10^{-3}} = 129{,}000$$
$$= 1.29 \times 10^5 \text{ colony-forming units/} \atop \text{ml or gram of food}$$

exercise 54

Microbes in Food: Contamination

Name _____

Date _____

Lab Section _____

Purpose _____

_____

_____

Data

Sample: _____

Dilution	Colonies per Plate
1:	
1:	
1:	
1:	

Number of colony-forming units per ml (or gram) of original food: _____

Do your calculations in the space below.

Sample: _____

Dilution	Colonies per Plate
1:	
1:	
1:	
1:	

Number of colony-forming units per ml (or gram) of original food: _____

Do your calculations in the space below.

Record data for other foods tested by other students.

Food	cfu/ml (or gram)

Questions

1. What could you do to ensure that the bacteria present in foods do not pose a health hazard? _____

    _____

    _____

2. Why are plates with 25 to 250 colonies used for calculations? _____

    _____

    _____

3. In a quality-control laboratory, each dilution is plated in duplicate or triplicate. Why would this increase the

    accuracy of a standard plate count? _____

    _____

    _____

4. There are other techniques for counting bacteria, such as a direct microscopic count and turbidity. Why is the

    standard plate count preferred for food? _____

    _____

    _____

5. Why is ground beef a better bacterial growth medium than a steak or roast? _____

    _____

6. Why does repeated freezing and thawing increase bacterial growth in meat? _____

    _____

Critical Thinking

1. Are bacteria identified in the standard plate count? _____ Why is a standard plate count performed on

    food? _____

2. Pasteurized milk is allowed 20,000 cfu/ml. How many colonies would be present if 1 ml of a 10^{-3} dilution was

    plated? _____ Is 20,000 cfu/ml a health hazard? _____

3. An outbreak of botulism associated with home-canned chili peppers killed 16 members of one family. Would a
    standard plate count have detected the etiologic agent in the canned chili peppers? Briefly explain.

exercise 55

Microbes Used in the Production of Foods

Objectives

After completing this exercise, you should be able to:

1. Explain how the activities of microorganisms are used to preserve food.
2. Define fermentation.
3. Produce an enjoyable product.

Background

Microbial fermentations are used to produce a wide variety of foods. **Fermentation** means different things to different people. In industrial usage, it is any large-scale microbial process occurring with or without air. To a biochemist, it is the group of metabolic processes that release energy from a sugar or other organic molecule, do not require oxygen or an electron transport system, and use an organic molecule as a final electron acceptor. In this exercise, we will examine a lactic acid fermentation used in the production of food.

In dairy fermentations, such as yogurt production, microorganisms use lactose and produce lactic acid without using oxygen. In nondairy fermentations, such as wine production, yeast use sucrose to produce ethyl alcohol and carbon dioxide under anaerobic conditions. If oxygen is available, the yeast will grow aerobically, liberating carbon dioxide and water as metabolic end-products.

Historically, milk has been fermented by selected microbes, with the resulting acidity preventing spoilage by acid-intolerant microbes. These "sour" milks have varied from country to country depending on the source of milk, conditions of culture, and microbial "starter" used. Milk from donkeys to zebras has been used, with the Russian *kumiss* (horse milk), containing 2% alcohol, and Swedish *surmjölk* (reindeer milk) being unusual examples. The bacteria yield lactic acid, and the yeast produce ethyl alcohol. Currently, two fermented cow milk products, buttermilk and yogurt, are widely used.

Buttermilk is the fluid left after cream is churned into butter. Today, buttermilk is actually prepared by souring true buttermilk or by adding bacteria to skim milk and then flavoring it with butterflake. *Lactococcus lactis* ferments the milk, producing lactic acid (sour); and neutral fermentation products (diacetyls) are produced by *Leuconostoc*. Yogurt originated in the Balkan countries, goat milk being the primary source. Yogurt is milk that has been concentrated by heating and then fermented at elevated temperatures. *Streptococcus* produces lactic acid, and *Lactobacillus* produces the flavors and aroma of yogurt.

Materials

Homogenized milk

Nonfat dry milk

Large beaker

Stirring rod

Thermometer

Glass test tube

Hot plate or ring stand and asbestos pad

5 ml pipette

Styrofoam or paper cups with lids

Second period:

Plastic spoons

Petri plate containing trypticase soy agar

pH paper

Gram stain reagents

Optional: jam, jelly, honey, and so on

Culture

Commercial yogurt or *Streptococcus thermophilus* and *Lactobacillus bulgaricus*

Techniques Required

Gram staining, Exercise 5

Plate streaking, Exercise 11

Pipetting, Appendix A

Procedure

Be sure all glassware is clean.

1. Add 100 ml of milk per person (in your group) to a wet beaker (wash out the beaker first with water to decrease the sticking of the milk). Put a thermometer in a glass test tube containing water before placing it in the beaker.

2. Heat the milk on a hot plate or over a burner on an asbestos pad placed on a ring stand, to about 80°C for 10 to 20 minutes. Stir occasionally. Do not let it boil. Why is the milk heated? _____
_____

3. Cool the milk to about 65°C and add 3 g of nonfat dry milk per person. Stir to dissolve. Why is dry milk added? _____
_____

4. Rapidly cool the milk to about 45°C. Pour the milk equally into the cups.

5. Inoculate each cup with 1 to 2 teaspoonfuls of commercial yogurt, or 2.5 ml *S. thermophilus* and 2.5 ml *L. bulgaricus.* Cover and label the cups.

6. Incubate the cups at 45°C for 4 to 18 hours or until they are firm (custardlike).

7. Cool the yogurt to about 5°C. Save a small amount for steps 8 through 10. Then taste it with a clean spoon. Add jam or some other flavor if you desire. Eat!

 Do not work with other bacteria or perform other exercises while eating the yogurt.

---

8. Determine the pH of the yogurt.

9. Make a smear, and after heat-fixing it, Gram stain it. Record your results.

10. Streak for isolation on trypticase soy agar. Incubate the plate, inverted, at 45°C.

11. After distinct colonies are visible, record your observations. Prepare Gram stains from each different colony.

exercise 55

Microbes Used in the Production of Foods

Name _____

Date _____

Lab Section _____

Purpose _____

_____

_____

Observations

Yogurt characteristics:

  Taste: _____

  Consistency: _____

  Odor: _____

  pH: _____

Gram stain results: _____

_____

Streak plate results: _____

_____

_____

Gram stain of isolated colonies: _____

_____

Questions

1. How can pathogens enter yogurt, and how can this be prevented? _____

_____

_____

2. What could cause an inferior product in a microbial fermentation process? _____

_____

3. How are microbial fermentations used to preserve foods? _____

_____

4. When the yogurt is Gram stained, what result would indicate that the yogurt is contaminated? _____

_____

Critical Thinking

1. What was the source of the bacteria and yeast originally used in dairy product fermentations and breads?

2. Describe how you would show that *Streptococcus* and *Lactobacillus* are in commercial yogurt.

3. How would consumption of yogurt on a regular basis affect the indigenous microbiota of the intestinal tract?

exercise 56

Microbes in Soil: The Nitrogen and Sulfur Cycles

Soil is the placenta of life.

PETER FARB

Objectives

After completing this exercise, you should be able to:

1. Diagram the nitrogen cycle, showing the chemical changes that occur at each step.
2. Explain the importance of the nitrogen cycle.
3. Differentiate between symbiotic and nonsymbiotic nitrogen fixation.
4. Diagram the sulfur cycle as it occurs in a Winogradsky column.

Background

One aspect of soil microbiology that has been studied extensively is the nitrogen cycle. All organisms need nitrogen for the synthesis of proteins, nucleic acids, and other nitrogen-containing compounds. The recycling of nitrogen by different organisms is called the **nitrogen cycle** (Figure 56.1). Microbes play a fundamental, irreplaceable role in the nitrogen cycle by participating in many different metabolic reactions involving nitrogen-containing compounds. When plants, animals, and microorganisms die, microbes decompose them by proteolysis and ammonification.

Proteolysis is the hydrolysis of proteins to form amino acids. **Ammonification** liberates ammonia by deamination of amino acids or catabolism of urea to ammonia. In most soil, ammonia dissolves in water to form ammonium ions:

$$NH_3 + H_2O \rightarrow NH_4OH \rightarrow NH_4^+ + OH^-$$

Ammonia Water Ammonium   Ammonium Hydroxyl
                 hydroxide   ion   ion

Some of the ammonium ions are used directly by plants and bacteria for the synthesis of amino acids.

The next sequence of the nitrogen cycle is the oxidation of ammonium ions in **nitrification.** Two genera of soil bacteria are capable of oxidizing ammonium ion in the two successive stages shown as follows:

$$2 NH_4^+ + 2 O_2 \xrightarrow{Nitrosomonas} 2 NO_2^-$$

Ammonium   Oxygen   Nitrite ion
ion

$$2 NO_2^- + O_2 \xrightarrow{Nitrobacter} 2 NO_3^-$$

Nitrite ions Oxygen   Nitrate ions

These reactions are used to generate energy (ATP) for the cells. Nitrifying bacteria are chemoautotrophs, and many are inhibited by organic matter. Nitrates are an important source of nitrogen for plants.

Denitrifying bacteria reduce nitrates and remove them from the nitrogen cycle. **Denitrification** is the reduction of nitrates to nitrites and nitrogen gas. This conversion may be represented as follows:

$$NO_3^- \rightarrow NO_2^- \rightarrow N_2O \rightarrow N_2$$

Nitrate Nitrite Nitrous Nitrogen
ion   ion   oxide   gas

Denitrification is **anaerobic respiration.** Many genera of bacteria, including *Pseudomonas* and *Bacillus,* are capable of denitrification under anaerobic conditions.

Atmospheric nitrogen can be returned to the soil by the conversion of nitrogen gas into ammonia, a process called **nitrogen fixation.** Microbes possessing the nitrogenase enzyme can fix nitrogen under anaerobic conditions as follows:

$$N_2 + 6 H^+ + 6 e^- \longrightarrow 2 NH_3$$

Nitrogen Hydrogen Electrons   Ammonia
gas   ions

Some free-living prokaryotic organisms, such as *Azotobacter,* clostridia, and cyanobacteria, can fix nitrogen. Many of the nitrogen-fixing bacteria live in close association with the roots of grasses in the **rhizosphere,** where root hairs contact the soil.

Symbiotic bacteria serve a more important role in nitrogen fixation. One such symbiotic relationship is a

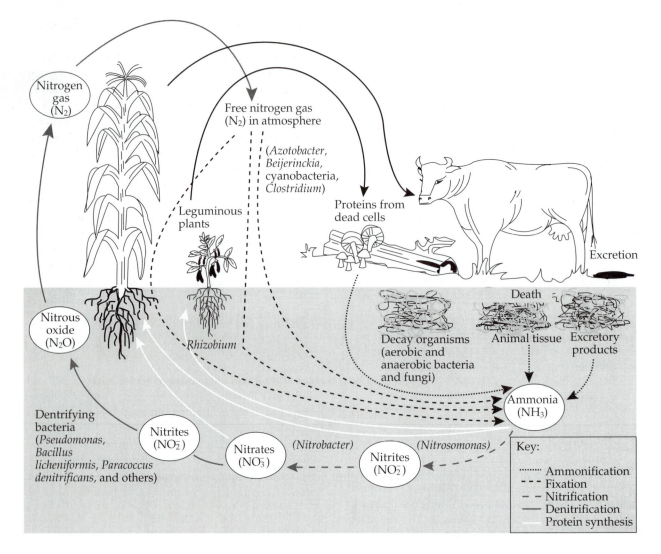

Figure 56.1
The nitrogen cycle.

mutualistic relationship between *Rhizobium* and the roots of legumes (such as soybeans, beans, peas, alfalfa, and clover) (Figure 56.2). There are thousands of legumes. Farmers grow soybeans and alfalfa to replace nitrogen in their fields. Many wild legumes are able to grow in the poor soils found in tropical rain forests or arid deserts. *Rhizobium* species are specific for the host legume that they infect. When a root hair and rhizobia make contact in the soil, a root nodule forms on the plant. The nodule provides the anaerobic environment necessary for nitrogen fixation.

Symbiotic nitrogen fixation also occurs in the roots of nonleguminous plants. The actinomycete *Frankia* forms root nodules in alders.

Any break in the nitrogen cycle would be critical to the survival of all life.

Purple and green bacteria are involved in another biogeochemical cycle, the **sulfur cycle** (Figure 56.3).

Green photosynthetic bacteria are colored by bacteriochlorophylls although they may appear brown due to the presence of red accessory photosynthetic pigments called **carotenoids. Purple photosynthetic bacteria** appear purple or red because of large amounts of carotenoids. Purple bacteria also have bacteriochlorophylls.

Photosynthetic bacteria use **bacteriochlorophylls** to generate electrons for ATP synthesis and use *sulfur, sulfur-containing compounds, hydrogen gas,* or *organic*

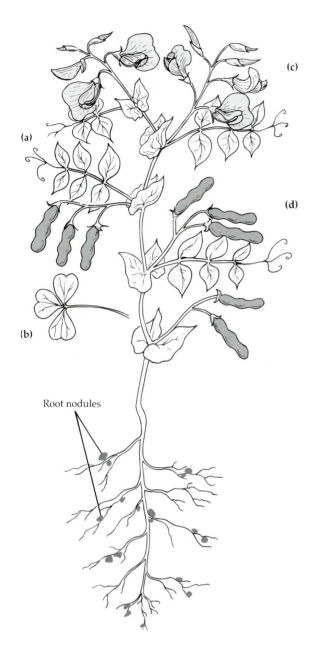

Figure 56.2
Some characteristic features of legumes. The leaves are opposite and may be **(a)** pinnately or **(b)** palmately compound. The flowers **(c)** have five asymmetrical petals, and the fruits **(d)** are pealike.

Root nodules

molecules as electron donors. The generalized equation for bacterial photosynthesis is as follows:

$$6\,CO_2 \qquad + \qquad 12\,H_2S$$
Carbon dioxide   Hydrogen sulfide

$$\xrightarrow{\text{Bacteriochlorophyll}} C_6H_{12}O_6 + \quad 12\,S^{2-} \quad + 6\,H_2O$$
Sugar   Sulfide ion   Water

Some photosynthetic bacteria store sulfur granules in or on their cells as a result of the production of sulfide ions. The stored sulfur can be used as an electron do-

nor in photosynthesis, resulting in the production of sulfates.

In nature, hydrogen sulfide is produced from the reduction of sulfates in anaerobic respiration and the degradation of sulfur-containing amino acids. Sulfates can be reduced to hydrogen sulfide by five genera of *sulfate-reducing* bacteria (the best known of which is *Desulfovibrio*). Carbon dioxide used by photosynthetic bacteria is provided by the fermentation of carbohydrates in an anaerobic environment.

In this exercise, we will use an enrichment culture technique involving a habitat-simulating device called a Winogradsky column to enhance the growth of bacteria involved in the anaerobic sulfur cycle. A variety of organisms will be cultured depending on their exposure to light and the availability of oxygen (Figure 56.4). We will also investigate the steps in the nitrogen cycle.

Materials

Ammonification
Tube containing peptone broth

Soil (Bring your own.)

Second period:

Nessler's reagent

Tube containing peptone broth

Ammonium hydroxide

Spot plate

Denitrification
Tubes containing nitrate-salts broth (2)

Soil (Bring your own.)

Second period:

Nitrate reagents A and B

Zinc dust

Nitrogen Fixation
Petri plate containing mannitol–yeast extract agar

Methylene blue

Sterile razor blade

Legumes (Bring your own; see Figure 56.2.)

Winogradsky Column
Mud mixture (mud, $CaCO_3$, hay or paper, and $CaSO_4$)

Large test tube or graduated cylinder

Glass rod

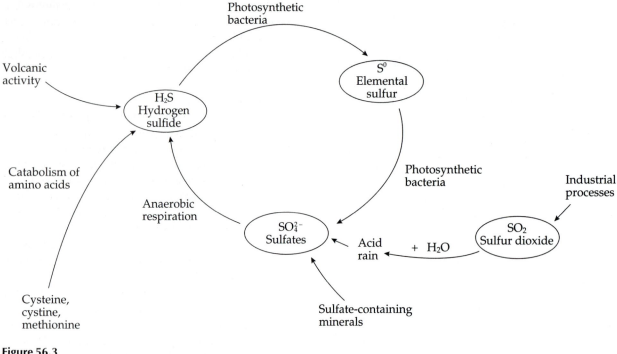

Figure 56.3
Anaerobic sulfur cycle.

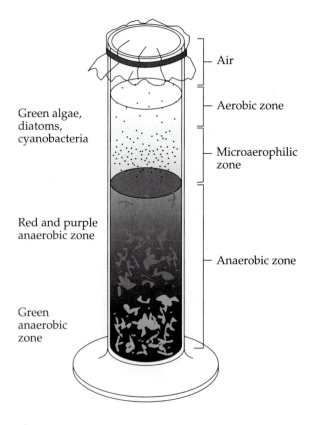

Figure 56.4
Winogradsky column.

Plain mud

Winogradsky buffer (NH_4Cl, Na_2S, KH_2PO_4, K_2HPO_4)

Aluminum foil

Light source

Culture

Pseudomonas aeruginosa

Demonstrations

Rhizobium inoculum used by farmers

Slides of stained root nodules

Techniques Required

Compound light microscopy, Exercise 1

Wet mount, Exercise 2

Simple staining, Exercise 3

Aseptic technique, Exercise 10

Nitrate reduction test, Exercise 17

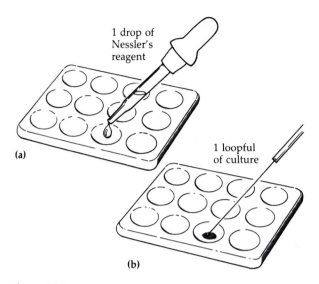

Figure 56.5
(a) Place a drop of Nessler's reagent in a well of a spot plate. **(b)** Add a loopful of the sample to be tested, and mix. Observe for a color change.

Procedure

Ammonification

1. Obtain an inoculum of soil on a wetted loop and inoculate the peptone broth. What is peptone? ___

2. Incubate the peptone broth at room temperature and test for ammonia at 2 days and at 7 days.

3. Test for ammonia by placing a drop of Nessler's reagent in a spot plate. Add a loopful of peptone broth, and mix (Figure 56.5). A yellow to brown color indicates the presence of ammonia. Compare your results to a spot to which a drop of ammonium hydroxide and Nessler's reagent have been added. Use an uninoculated tube of peptone broth as a control. Why is a control necessary? _____

Denitrification

1. Inoculate one nitrate-salts broth with soil using a wetted loop to pick up the soil. Inoculate the other tube with *P. aeruginosa*.

2. Incubate both tubes at room temperature for 1 week.

3. Test for nitrate reduction. Remember how? Describe. _____

Record your results.

Nitrogen Fixation

1. Cut off a nodule from a legume and wash it under tap water. (See color plate VI.4.) Observe.

2. Cut the nodule in half with a sterile razor blade. Observe. Crush the nodule between two slides and make a smear by rotating the slides together.

3. Streak a loopful of the crushed nodule on the mannitol–yeast extract agar. Incubate the plate at room temperature for 7 days.

4. Air dry the slide and heat-fix it. Stain the slide for 1 minute with methylene blue. Rinse the slide and observe it under oil immersion.

5. Observe the demonstrations. What is in the farmer's inoculum? _____

6. Observe the growth on the plate, and make a simple stain of the growth with methylene blue. Compare this stain to the stain prepared from the nodule.

Winogradsky Column

1. Pack a large test tube two-thirds full with the mud mixture. Pack it to eliminate air bubbles. Why? __

2. Carefully pack a narrow layer of plain mud on top of the first layer.

3. Gently pour the buffer down the side of the tube, taking care not to disturb the mud surface. Fill the tube as full as possible.

4. Cover the top of the tube with foil, and place it in front of a light source. The instructor may assign different light sources (such as incandescent, fluorescent, red, or green).

5. Observe the tubes at weekly intervals for 4 weeks. Record the appearance of colored areas. Aerobic mud will be brownish, and anaerobic mud will be black. Why? _____

6. After 4 weeks, prepare wet mounts from the purple or green patches. Record the microscopic appearance of the bacteria and whether sulfur granules are present (see color plate VI.3).

exercise 56

Microbes in Soil: The Nitrogen and Sulfur Cycles

Name _____

Date _____

Lab Section _____

Purpose _____

_____

_____

Data

Ammonification

Incubation	Growth	Nessler's Reagent			Ammonia Present?
		Color of Inoculated Peptone	Color of NH$_4$OH	Color of Control	
2 days					
7 days					

Denitrification

How did you test for denitrification? _____

Inoculum	Growth	Gas	Nitrate Reduction?	Denitrification?
Soil				
Pseudomonas				

Nitrogen Fixation

Diagram of root nodule:

Diagram microscopic appearance of cells in root nodule.

Morphology: _____

Diagram microscopic appearance of bacteria from culture.

Demonstration slides:
Label root, nodule, and bacteria.

100×

Morphology: _____

1000×

Describe colonies on the mannitol-yeast extract plate. _____
_____
_____

Winogradsky Column

Diagram the appearance of your enrichment at weekly intervals. Label patches of photosynthetic bacteria.

Light source: _____

| 1 Week | 2 Weeks | 3 Weeks | 4 Weeks |

Microscopic observations:

Source	Morphology	Motility	Sulfur Granules Present?

If other light sources were used, compare the appearance of the enrichments after 4 weeks. _____

_____

_____

_____

Questions

1. Write the equation for ammonification of any amino acid.

2. Why can't you smell ammonia in the peptone broth tubes? _____

_____

_____

3. Why is denitrification a problem for farmers? _____

    _____

    _____

4. What gas is in the gas tube of a positive denitrification test? _____

    _____

    _____

5. Which step(s) in the nitrogen cycle require oxygen? _____

    _____

    _____

6. Why is the nitrogen cycle important to all forms of life? _____

    _____

    _____

    _____

7. What was the purpose of each of the following chemicals in the Winogradsky enrichment?

    $CaCO_3$ _____ Na_2S _____

    Hay or paper _____ KH_2PO_4 and K_2HPO_4 _____

    $CaSO_4$ _____ NH_4Cl _____

8. Indicate the aerobic and anaerobic regions on the diagrams of your enrichment column. How can you tell? ____

    _____

9. Is there evidence of any nonphotosynthetic growth in the Winogradsky column? _____

    Explain. _____

    _____

Critical Thinking

1. What genetic engineering project would you propose concerning nitrogen fixation?

2. Why do the bacteria in the root nodule look different from the bacteria cultured from the nodule?

3. Design an experiment that would determine the effect of heat produced by the lights on the growth of bacteria in the Winogradsky column.

exercise 57

Microbes in Soil: Bioremediation

Objectives

After completing this exercise, you should be able to:

1. Define the following terms: bioremediation, lipid, hydrocarbon.
2. Demonstrate enrichment of oil-degrading bacteria.
3. Describe the digestion of lipids.

Background

The use of bacteria to eliminate pollutants is called **bioremediation.** Unlike some forms of environmental cleanup, in which dangerous substances are removed from one place only to be dumped in another, bacterial cleanup eliminates the toxic substance and often returns a harmless or useful substance to the environment. (See color plate X.1.)

Many bacteria use exoenzymes such as lipase to hydrolytically decompose lipids. The lipid molecule is first broken down into glycerol and three fatty acids (Figure 57.1). Some bacteria ferment the glycerol, while others oxidize the fatty acids. In the presence of air, the bacteria remove two carbons at a time, in a process called *beta oxidation*. When lipid hydrolysis occurs in decomposition of foods such as butter, it results in a rancid flavor and aroma due to the fatty acids. Bacterial cultures are being used to degrade grease from restaurants and meat processors before disposal. On their own, bacteria degrade large, complex molecules in petroleum too slowly to be helpful in cleaning up an oil spill. However, scientists have hit upon a very simple way to speed them up. Nitrogen and phosphorous plant fertilizers can be added, in a process called *bioenhancement.* This approach was used on an Alaskan beach affected by the 1989 *Valdez* oil spill; as a result, the beach is now free of oil.

In this exercise, we will enrich for hydrocarbon-degrading bacteria. Degradation of the hydrocarbon should result in the opaque, emulsified oil becoming small, soluble molecules, such as fatty acids and acetyl groups. (See color plate X.2.)

Materials

Soil slurry

Enrichment (hydrocarbon–minimal salts broth inoculated with soil slurry and incubated)

Sterile 1 ml pipettes (6)

Propipette or pipette bulb

Sterile 99 ml dilution blanks (4)

Petri plates containing hydrocarbon–minimal salts (4)

Spreading rod

Figure 57.1
Tributyrin digestion by hydrolysis. Lipids are composed of a glycerol and three fatty acid molecules.

Alcohol

Gram staining reagents

Techniques Required

Compound light microscopy, Exercise 1

Gram staining, Exercise 5

Enrichment techniques, Exercise 12

Spreading rod, Exercise 27

Pipetting, Appendix A

Serial dilution technique, Appendix B

Procedure

First Period

1. Obtain a soil slurry.
2. Using a sterile 1 ml pipette, aseptically transfer 1 ml of the soil slurry to a 99 ml dilution blank; label the bottle "1:10^2" and discard the pipette. Mix the contents of the bottle by shaking it 20 times.
3. Make a 1:10^4 dilution by transferring 1 ml of the 1:10^2 dilution to another 99 ml dilution blank. Thoroughly mix the contents of the bottle as before.
4. Label the bottoms of two hydrocarbon agar plates "1 ml" and "0.1 ml." Which hydrocarbon is in the plates? _____

5. Using a 1 ml pipette, aseptically transfer 0.1 ml of the 1:10^4 dilution onto the agar of the 0.1 ml plate. Using the same pipette, transfer 1 ml of the 1:10^4 dilution onto the agar of the 1 ml plate.

> ⚠️ **Disinfect the spreading rod by dipping it in alcohol. Keep the flame away from the beaker.**

Spread the inoculum over the agar using the rod. Incubate the plate at room temperature for 24 to 48 hours. (See color plate XII.5.)
6. Obtain an enrichment of the same hydrocarbon used in the plates in step 4. What is the hydrocarbon? _____
7. Repeat steps 2 through 5 to inoculate these plates from the enrichment.

Second Period

1. Observe and describe the growth on the plates. Count the number of hydrocarbon-degrading colonies. How will you identify them? _____ Calculate the number of hydrocarbon-degrading bacteria using the following formula:

$$\frac{\text{Number of colonies}}{\text{Amount plated} \times \text{dilution}} = \frac{\text{Colony-forming units/}}{\text{gram of soil}}$$

2. Gram stain some of the oil-degrading colonies.

exercise 57

Microbes in Soil: Bioremediation

Name _____

Date _____

Lab Section _____

Purpose _____

_____

_____

Data

Before enrichment:

Color Description	Oil Degrader?	Morphology and Gram Reaction	Number of This Type

Number of hydrocarbon-degrading cfu/gram of soil: _____

After enrichment:

Color Description	Oil Degrader?	Morphology and Gram Reaction	Number of This Type

Number of hydrocarbon-degrading cfu/gram of soil: _____

Conclusions

1. Did you isolate any oil-degrading bacteria? How can you tell? _____

   _____

   _____

2. Compare the results before and after enrichment. _____

   _____

   _____

Questions

1. What is an enrichment? _____

2. Besides clearing, how else could you determine whether lipids had been hydrolyzed? _____

   _____

   _____

3. The enrichment broth consisted of a hydrocarbon and minimal salts. What was the purpose of the hydrocarbon?

   _____

   Of the minimal salts? _____

   _____

Critical Thinking

1. Show beta oxidation of the following hydrocarbon:

$$H-\overset{\overset{\displaystyle H}{|}}{\underset{\underset{\displaystyle H}{|}}{C}}-\overset{\overset{\displaystyle H}{|}}{\underset{\underset{\displaystyle H}{|}}{C}}-\overset{\overset{\displaystyle H}{|}}{\underset{\underset{\displaystyle H}{|}}{C}}-\overset{\overset{\displaystyle H}{|}}{\underset{\underset{\displaystyle H}{|}}{C}}-\overset{\overset{\displaystyle H}{|}}{\underset{\underset{\displaystyle H}{|}}{C}}-\overset{\overset{\displaystyle H}{|}}{\underset{\underset{\displaystyle H}{|}}{C}}-\overset{\overset{\displaystyle H}{|}}{\underset{\underset{\displaystyle H}{|}}{C}}-\overset{\overset{\displaystyle H}{|}}{\underset{\underset{\displaystyle H}{|}}{C}}-\overset{\overset{\displaystyle H}{|}}{\underset{\underset{\displaystyle H}{|}}{C}}-\overset{\overset{\displaystyle H}{|}}{\underset{\underset{\displaystyle H}{|}}{C}}-H$$

2. Here are the partial formulas of two detergents that have been manufactured:

   Which of these would be readily degraded by bacteria? Why?

3.   How could the oil-degrading bacteria you grew be used commercially in a detergent or drain cleaner?

4.   Design an experiment to isolate mercury-utilizing bacteria from soil.

Appendices

EXERCISES

A Pipetting
B Dilution Techniques and Calculations
C Use of the Spectrophotometer
D Graphing
E Use of the Dissecting Microscope
F Use of the Membrane Filter
G Electrophoresis
H Keys to Bacteria

---

appendix A

Pipetting

Pipettes are used for measuring small volumes of fluid. They are usually coded according to the total volume and graduation units (Figure A.1). To fill a pipette, use a bulb or other mechanical device, as shown in Figure A.2.

> ☣ **Never use your mouth to fill a pipette.**

Draw the desired amount of fluid into the pipette. Read the volume at the bottom of the meniscus (Figure A.3).

Figure A.1
This pipette holds a total volume of 1 milliliter when filled to the zero mark. It is graduated in 0.01 ml units.

(a)   (b)

Figure A.2
Two types of pipette aspirators. **(a)** Attach this plastic pump to the pipette, and turn the wheel to draw fluid into the pipette. Turning the wheel in the other direction will release the fluid. **(b)** Insert a pipette into this bulb. While pressing the A valve, squeeze the bulb, and it will remain collapsed. To draw fluid into the pipette, press the S valve; to release fluid, press the E valve.

— Meniscus

Figure A.3
Read the fluid volume at the lowest level of the meniscus.

(a)

(b)

Figure A.4
(a) A serological pipette. **(b)** A measuring pipette.

Fill the pipette to above the zero mark, then allow it to drain just to the zero. The desired amount can then be dispensed.

Microbiologists use two types of pipettes. The **serological pipette** is meant to be emptied to deliver the total volume (Figure A.4a). Note that the graduations stop above the tip. The **measuring pipette** delivers the volume read on the graduations (Figure A.4b). This pipette is not emptied, but the flow must be stopped when the meniscus reaches the desired level.

Aseptic use of a pipette is often required in microbiology. Bring the entire closed pipette container to your work area. If the pipettes are wrapped in paper, open the wrapper at the end opposite the delivery end; in a canister, the delivery end will be at the bottom of the canister. Do not touch the delivery end of a sterile pipette. Remove a pipette, attach the pipette aspirator, and with your other hand, pick up the sample to be pipetted. Remove the cap from the sample with the little finger of the hand holding the pipette. Fill the pipette, and replace the cap on the sample.

> 🔬 **After pipetting, place the contaminated pipette in the appropriate container of disinfectant. If it is a disposable pipette, discard it in a biohazard bag or any container designated for biohazards.**

Micropipettes are used to measure volumes less than 0.5 ml. The design and operation of micropipettes varies according to the manufacturer. Most use a disposable tip to hold the fluid, as shown in Figure A.5. To use a micropipette, select one that holds the volume you need. The micropipette is labeled with its range—for example, 0.5–10 µl, 1–10 µl, 10–100 µl, 100–1000 µl. Set the desired volume on the digital display (Figure A.6a) by turning the control knob (Figure A.6b).

Questions

1. Which micropipette would you use to measure

   15 µl? _____

2. If this is the display for the micropipette you chose for question 1, write in the 1 and 5 to show 15 µl.

   (a)   (b)   (c)   (d)

Figure A.5

Generalized use of a micropipette. (**a**) Depress the control button, and insert the tip into the liquid. (**b**) Smoothly release the button to allow the liquid to enter the tip. (**c**) Place the tip against the inside of the receiving tube, and depress the button. (**d**) Eject the used tip by pressing the eject button or by pressing the control button to the final stop.

3. How do you eject the tip on your micropipette?

   _____

(a)

(b)

Figure A.6

Adjusting a micropipette. (**a**) Reading the display. (**b**) Turning the control knob on this micropipette changes the volume picked up when the control knob is depressed.

appendix B

*Dilution Techniques and Calculations**

Bacteria, under good growing conditions, will multiply into such large populations that it is often necessary to dilute them to isolate single colonies or to obtain estimates of their numbers. This requires mixing a small, accurately measured sample with a large volume of sterile water or saline, which is called the **diluent** or **dilution blank.** Accurate dilutions of a sample are obtained through the use of pipettes. For convenience, dilutions are usually made in multiples of ten.

A single dilution is calculated as follows:

$$\text{Dilution} = \frac{\text{Volume of the sample}}{\text{Total volume of the sample + the diluent}}$$

For example, the dilution of 1 milliliter into 9 milliliters equals

$$\frac{1}{1 + 9}, \text{ which is } \frac{1}{10} \text{ and is written 1:10}$$

The same formula applies for all dilutions, regardless of the volumes. A dilution of 0.1 milliliter into 0.9 milliliter equals

$$\frac{0.1}{0.1 + 0.9}, \text{ which is } \frac{0.1}{1} = \frac{1}{10} = 1:10$$

A dilution of 0.5 milliliter into 4.5 milliliters equals

$$\frac{0.5}{0.5 + 4.5}, \text{ which is } \frac{0.5}{5.0} = \frac{1}{10} = 1:10$$

*Adapted from C. W. Brady. "Dilutions and Dilution Calculations." Unpublished paper. Whitewater, WI: University of Wisconsin, n.d.

Experience has shown that better accuracy is obtained with very large dilutions if the total dilution is made out of a series of smaller dilutions rather than one large dilution. This series is called a **serial dilution,** and the total dilution is the product of each dilution in the series. For example, if 1 ml is diluted with 9 ml, and then 1 ml of that dilution is put into a second 9 ml diluent, the final dilution will be

$$\frac{1}{10} \times \frac{1}{10} = \frac{1}{100} \quad \text{or} \quad 1:100$$

To facilitate calculations, the dilution is written in exponential notation. In the example above, the final dilution 1:100 would be written 10^{-2}. Remember, $1:100 = \frac{1}{100} = 0.01 = 10^{-2}$. (See the section Exponents, Exponential Notation, and Logarithms, on page 414.) A serial dilution is illustrated in Figure B.1.

Procedure

1. Aseptically pipette 1 ml of sample into a dilution blank.
   a. If the dilution is into a tube, mix the contents by rolling the tube back and forth between your hands.
   b. If the dilution is into a 99 ml blank, hold the cap in place with your index finger and shake the bottle up and down through a 35 cm arc.

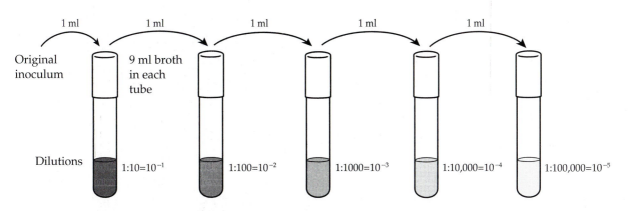

Figure B.1

A 1 ml sample from the first tube will contain ¹/₁₀ the number of cells present in 1 ml of the original sample. A 1 ml sample from the last tube will contain 1/100,000 the number of cells present in 1 ml of the original sample.

2. It is necessary to use a fresh pipette for each dilution in a series, but it is permissible to use the same pipette to remove several samples from the same bottle, as when plating out samples from a series of dilutions.

Problems

Practice calculating serial dilutions using the following problems.

1.

2.

3. Design a serial dilution to achieve a final dilution of 10^{-8}.

4.

5.

Exponents, Exponential Notation, and Logarithms*

Very large and very small numbers—such as 4,650,000,000 and 0.00000032—are cumbersome to work with. It is more convenient to express such numbers in exponential notation— that is, as a power of 10. For example, 4.65×10^9 is written in **standard exponential notation,** or **scientific notation;** 4.65 is the **coefficient,** and 9 is the power, or **exponent.** In standard exponential notation, the coefficient is a number between 1 and 10, and the exponent is a positive or negative number.

To change a number into exponential notation, follow two steps. First, determine the coefficient by moving the decimal point so you leave only one nonzero digit to the left of it. For example:

$$0.00000032$$

The coefficient is 3.2. Second, determine the exponent by counting the number of places you moved the decimal point. If you moved it to the left, the exponent is a positive number. If you moved it to the right, the exponent is negative. In the example, you moved the decimal point 7 places to the right, so the exponent is -7. Thus

$$0.00000032 = 3.2 \times 10^{-7}$$

Now suppose we are working with a very large number instead of a very small number. The same rules apply, but our exponential value will be positive rather than negative. For example:

$$4,650,000,000 = 4.65 \times 10^{+9}$$
$$= 4.65 \times 10^9$$

To multiply numbers written in exponential notation, multiply the coefficients and *add* the exponents. For example:

$$(3 \times 10^4) \times (2 \times 10^3) =$$
$$(3 \times 2) \times 10^{4+3} = 6 \times 10^7$$

To divide numbers written in exponential notation, divide the coefficients and *subtract* the exponents. For example:

$$\frac{3 \times 10^4}{2 \times 10^3} = \frac{3}{2} \times 10^{4-3} = 1.5 \times 10^1$$

Microbiologists use exponential notation in many kinds of situations. For instance, exponential notation is used to describe the number of microorganisms in a population. Such numbers are often very large. Another application of exponential notation is to express concentrations of chemicals in a solution—chemicals such as media components, disinfectants, or antibiotics. Such numbers are often very small. Converting from one unit of measurement to another in the metric system requires multiplying or dividing by a power of 10, which is easiest to carry out in exponential notation.

A **logarithm** is the power to which a base number is raised to produce a given number. Usually we work with logarithms to the base 10, abbreviated **$\log_{10}$.** The first step in finding the $\log_{10}$ of a number is to write the number in standard exponential notation. If the coefficient is exactly 1, the $\log_{10}$ is simply equal to the exponent. For example:

$$\log_{10} 0.00001 = \log_{10}(1 \times 10^{-5})$$
$$= -5$$

If the coefficient is not 1, as is often the case, a calculator must be used to determine the logarithm.

Microbiologists use logs for pH calculations and for graphing the growth of microbial populations in culture.

Source: G. J. Tortora, B. R. Funke, and C. L. Case. *Microbiology: An Introduction,* 7th ed. San Francisco, CA: Benjamin/Cummings, 2001, Appendix D.

appendix C

Use of the Spectrophotometer

In a **spectrophotometer,** a beam of light is transmitted through a bacterial suspension to a photoelectric cell (Figure C.1). As bacterial numbers increase, the broth becomes more turbid, causing the light to scatter and allowing less light to reach the photoelectric cell. The change in light is registered on the instrument as **percentage of transmission,** or **%T** (the amount of light getting through the suspension) and **absorbance (Abs.)** (a value derived from the percentage of transmission). Absorbance is a logarithmic value and is used to plot bacterial growth on a graph.

Operation of the Spectronic 20 and Meter-Model Spectronic 21 Spectrophotometers (Figure C.2)

1. Turn on the power and allow the instrument to warm up for 15 minutes.
2. Set the wavelength for maximum absorption of the bacteria and minimal absorption of the culture medium.

Figure C.2
Bausch and Lomb Spectronic 20.

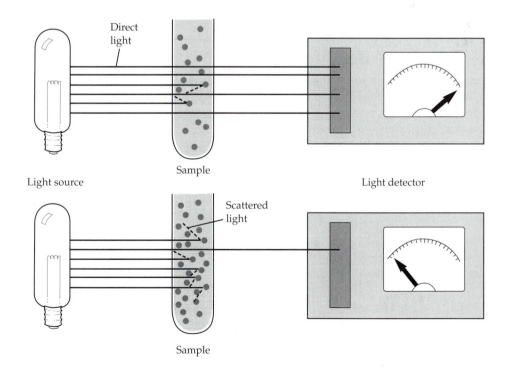

Figure C.1
Estimation of bacterial numbers by turbidity. The amount of light picked up by the light detector is proportional to the number of bacteria. The less light transmitted, the more bacteria in the sample.

3. For the Spectronic 20, *zero* the instrument by turning the zero control until the needle measures 0% transmission.
4. To read a metered scale, look directly at the meter so the needle is superimposed on its reflection in the mirror.
5. Place an uninoculated tube of culture medium (*control*) in the sample holder. To *standardize* the instrument, turn the light control until the needle registers 100% transmission.
6. To take a sample measurement, place an inoculated tube of culture medium in the sample holder and record the %T from the scale.
7. %T is usually read on the Spectronic 20 because it is the more accurate scale. Calculate the absorbance as follows:

$$\text{Absorbance} = -\log\left(\frac{\%\text{T}}{100}\right)$$

Operation of the Digital-Model Spectronic 21 (Figure C.3)

1. Turn on the power and allow the instrument to warm up for 15 minutes.
2. Set the wavelength for maximum absorption of the bacteria and minimal absorption of the culture medium.
3. Select the operating mode: A (absorbance) or %T (transmission). Absorbance data are more linear and usually preferable.
4. Set the sensitivity switch to LO.
5. Place an uninoculated tube of culture medium (*control*) in the sample holder. To *standardize* the instrument, turn the 100% T/Zero A control until the display registers 000 A. If 000 A can't be reached, turn the sensitivity to M or HI. A U in the

Figure C.3
Bausch and Lomb Spectronic 21, digital model.

upper-left corner of the display indicates that you are under the range of the light detector, and an O in the display indicates that you are over the range. In either case, use the sensitivity switch and the standardizing control to make the necessary adjustments.
6. To take a sample measurement, place an inoculated tube of culture medium in the sample holder and record the absorbance (Abs.) from the display.

appendix D

Graphing

A **graph** is a visual representation of the relationship between two variables. Whenever one variable changes in a definite way in relation to another variable, this relationship may be graphed.

Microbiologists work with large populations of bacteria and frequently use graphs to illustrate the activity of these populations. Graphs dealing with populations of cells are drawn on semi-log graph paper. The horizontal, or **X-axis,** is a linear scale. The X-axis is used for the **independent variable**—that is, the variable not being tested. The vertical, or **Y-axis,** is a logarithmic scale where each cycle represents a power of 10. The dependent variable is marked off along the Y-axis. The **dependent variable** changes in relation to the independent variable. The intersection point of the X-axis and Y-axis is called the **origin,** and all units are marked off from this point.

To draw a graph, first establish the two variables you want to graph—for example, time and number of bacteria (Table D.1). The independent variable is time, and the dependent variable is the number of bacteria. Mark the appropriate units of measure along the axes (Figure D.1).

We will graph the numbers of bacteria in the culture incubated at 20°C. The first point on the graph is $(0, 5.00 \times 10^5)$. Locate 0 on the X-axis, measure directly above it the distance 5.00×10^5, and mark this point. After a sufficient number of points have been plotted, a smooth curve can be drawn connecting these points. The line through the points should be straight and does not have to connect all the points. When the points do not fall in a straight line, draw a line between the points, leaving an equal number of points above and below the line.

Table D.1

Numbers of *E. coli* in Three Cultures Incubated at Different Temperatures

Time (hr)	Temperature (°C)		
	20	15	35
0	5.00×10^5	1.50×10^5	5.40×10^5
4	—	1.50×10^5	—
5	4.60×10^5	—	—
6	—	—	5.20×10^5
10	5.78×10^5	5.62×10^5	6.00×10^5
11	8.80×10^5	—	1.23×10^6
12	1.05×10^6	—	2.58×10^6
13	1.16×10^6	—	4.93×10^6
14	2.32×10^6	—	9.00×10^6
15	3.80×10^6	1.62×10^6	—
16	7.00×10^6	—	—
17	8.10×10^6	—	—
18	9.60×10^6	—	—
19	1.95×10^7	—	—
20	—	2.70×10^6	—

When graphs are drawn that compare the same variables under different conditions (for example, bacterial growth rates at different temperatures), it is best to draw all the graphs on the same paper so comparisons are obvious and easily made. Graphs should be titled to explain the data presented.

The data for cultures incubated at 15°C and 35°C (Table D.1) can be plotted on the graph in Figure D.1. It is not necessary to have bacterial numbers for the same "time points" in each graph. The missing points will be filled in by the line. This is possible because the line between each point is an interpretation of what was happening between measurements.

Activity

Using the data provided in Table D.1, graph the growth curves on the graph paper and compare the effect of temperature on the rate of growth. Since *rate* is a change over time, the rate can be interpreted by the slope of the line. What will you title this graph? _____

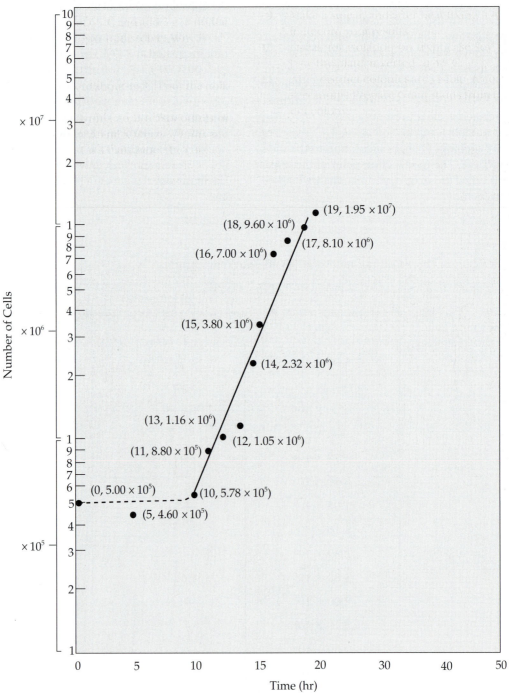

Figure D.1
Graph showing growth of *E. coli* at 20°C.

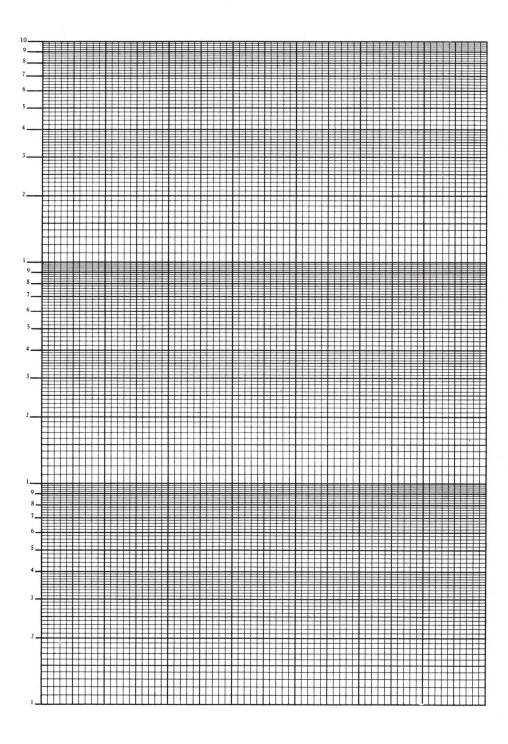

appendix E

Use of the Dissecting Microscope

The **dissecting microscope** (Figure E.1) is useful for observing specimens that are too large for the compound microscope and too small to be seen with the naked eye. The dissecting microscope is also called a **stereoscopic** microscope because the image is three-dimensional.

Dissecting microscopes are usually equipped with 10× ocular lenses and a 1× objective lens. Some dissecting microscopes have a rotating nosepiece that houses 1× and 2× objective lenses or a zoom objective lens (7×).

The dissecting microscope may have a built-in light source or a substage mirror and auxiliary lamp. Illumination can be adjusted on microscopes with two built-in lamps and/or by using a mirror. For transparent spec-imens, the light should be directed through the specimen from under the stage. For opaque specimens, light should be directed onto the specimen from the top. Both types of lighting can be used for observing a translucent specimen.

The specimen is placed on the stage. Some microscopes have an alternate black stage plate for use with opaque specimens. The stage clips can be moved to the side when large specimens, such as Petri plates, are used. While looking through the microscope, adjust the width of the eyepieces so you see one field of vision. If two circles or fields are visible, the ocular lenses are too far apart; if two fields overlap, the ocular lenses are too close together. To focus, turn the adjustment knob.

Figure E.1
The dissecting microscope: principal parts and their functions.

Ocular lenses Primary magnification, usually 10 times.

Nosepiece Houses objective lenses. May be rotated on some microscopes for additional magnification.

Objective lens Provides additional magnification.

Lamp Provides illumination.

Arm

Adjustment knob Moves lenses for focusing.

Mirror Directs light through specimen.

Stage clips May be used to hold specimen in place.

Stage Platform to hold specimen.

Stage plate Provides contrasting or transparent background for specimen.

Questions

1. What is the magnification of your dissecting

   microscope? _____

2. Using a ruler, measure the size of the field of vision

   at low power: _____ mm. At high power: _____
   mm.

3. How can you adjust the direction of illumination

   on your microscope? _____

   _____

appendix F

Use of the Membrane Filter

Membrane filtration can be used to separate bacteria, algae, yeasts, and molds from solutions. Nitrocellulose or polyvinyl membrane filters with 0.45 μm pores are commonly used to trap microorganisms. Membrane filtration is used to separate viruses from their host cells because the viruses will pass through the filter. Membrane filters are used to trap bacteria found in water in order to count the bacteria.

Membrane filters have many uses in industrial microbiology. They are used to filter wine, soft drinks, air, and water to detect microorganisms and particulate matter. In addition, membrane filters can be used to separate large molecules, such as DNA, from solutions.

The receiving flask, filter base, and filter support can be wrapped in paper and sterilized by autoclaving (Figure F.1). Membrane filters can also be sterilized by autoclaving. For use, the filter base must be unwrapped aseptically. With sterile, flat forceps, a sterile filter is

placed on the filter support, and the filter funnel is clamped or screwed into place.

Gravity alone will not pull a sample through the small filter pores, so the filtering flask is attached to a vacuum source (Figure F.1). Since air pressure is then greater outside the flask, the sample is pushed into the vacuum inside the flask. A vacuum pump or vacuum line is usually used to provide the necessary vacuum. An aspirator trap is placed between the vacuum source and the receiving flask to prevent the flow of solution into the pump or vacuum line. With small filters and small volumes of fluid, a syringe can be used to supply the needed vacuum.

Following filtration, the filtrate in the flask is free of all microorganisms larger than viruses. For observation of cells trapped on the filter, the filter can be dipped into immersion oil to make it translucent and then mounted on a microscope slide. Microorganisms can be cultured from the filter by placing the filter on a solid nutrient medium.

Figure F.1
Membrane filtration setup. (Plastic disposable filters can be substituted for glass filters.)

appendix G

Electrophoresis

Charged macromolecules can be separated according to size and charge by a process called **electrophoresis.** The material to be tested is placed at one end of a gel matrix. Polyacrylamide gel or agarose are used to separate proteins and small pieces of DNA and RNA. Agarose is the usual matrix for separation of nucleic acids. A buffer (salt solution) is used to conduct electric current through the matrix. When an electric current is applied, the molecules start to move. Molecules with different charges and sizes will migrate at different rates.

Procedure

1. Place the well-forming comb in position on a casting tray (Figure G.1). Seal the open ends of the casting tray with masking tape. Or, if the tray has movable gates, raise the gate at each end. Pour melted agarose onto the casting tray until the agar surrounds the teeth. The agar should be about 5 mm deep and have a flat surface. Let the gel harden for about 20 minutes. When the gel is solid, care-

Figure G.1
Preparation of an agarose minigel.

Electrophoresis chamber
(side view)

Electrophoresis chamber
(top view)

Figure G.2
Setup of electrophoresis equipment.

fully remove the comb and then remove the tape or lower the end gates.

2. Fill the buffer chamber with the appropriate buffer. Add just enough buffer to cover the gel. Place the agar in the buffer chamber with or without the casting tray.

3. Fill the wells with samples to be tested. Close the lid and apply current (Figure G.2).

4. During electrophoresis, a visible marker dye will migrate in front of the sample. If the dye isn't moving in the right direction, you have accidently reversed the polarity of the electric field. If the marker doesn't move and if there is little current, you have an open circuit or have not added the correct buffer. In this case, you must start over.

5. The two components of the tracking dye migrate at different rates. Why? _____

Turn off the power when the two dyes are 3 or 4 cm apart. Remove the gel. The gel can now be stained to locate the macromolecules. Store stained or unstained gels in sealed plastic bags in the refrigerator.

appendix H

Keys to Bacteria

Appendix H deals with the identification of bacteria. The information is arranged in **dichotomous keys.** In a dichotomous key, identification is based on successive questions, each of which has two possible answers (*di-* means two). After answering each question, the investigator is directed to another question until an organism is identified. There is no single "correct" way to write a key; the goal is to conclude with one unambiguous identification.

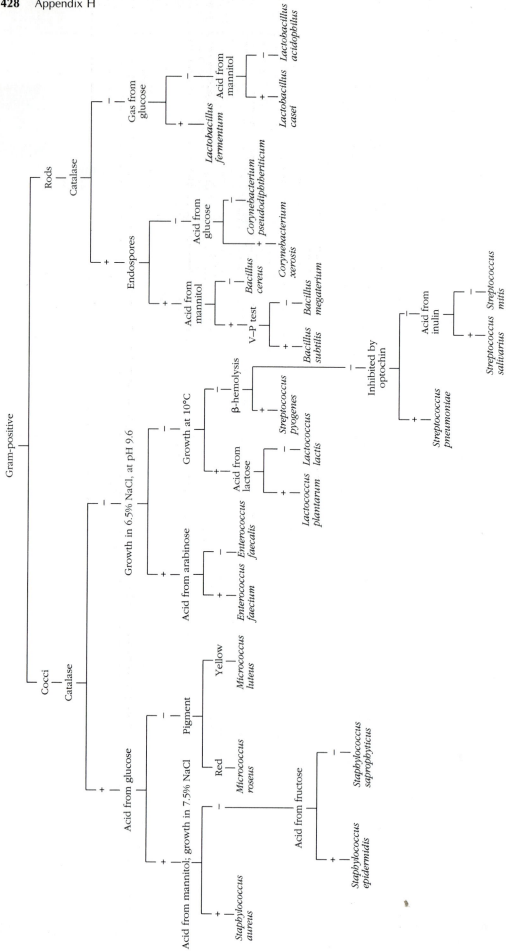

Figure H.1
Key to selected Gram-positive heterotrophic bacteria.

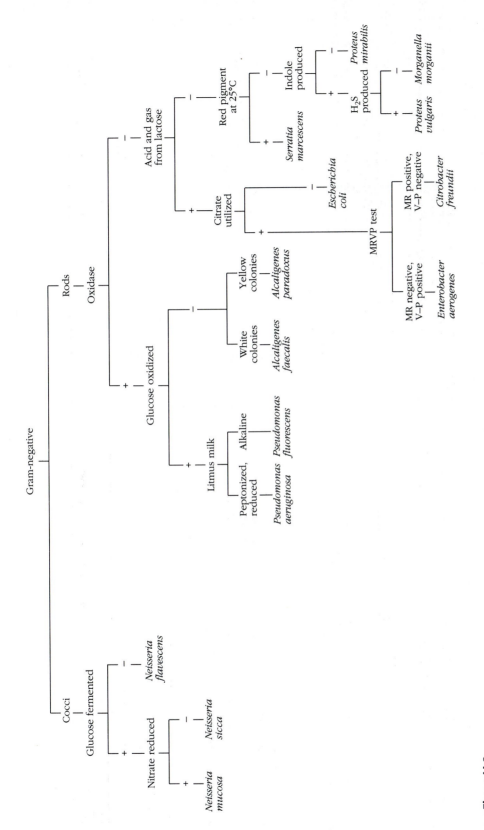

Figure H.2
Key to selected Gram-negative heterotrophic bacteria.

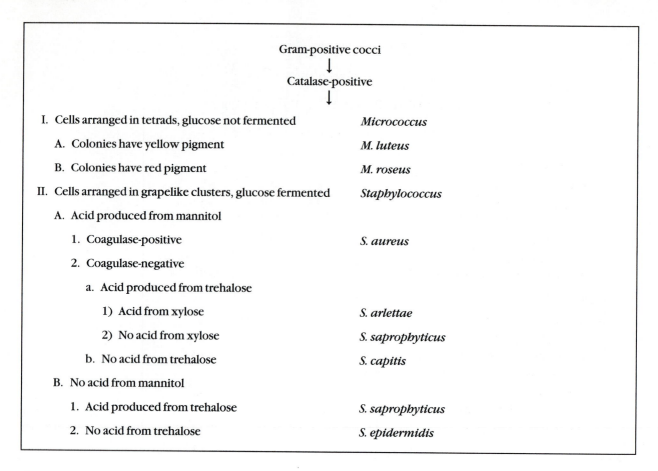

Figure H.3
Key to Gram-positive cocci commonly found on human skin.

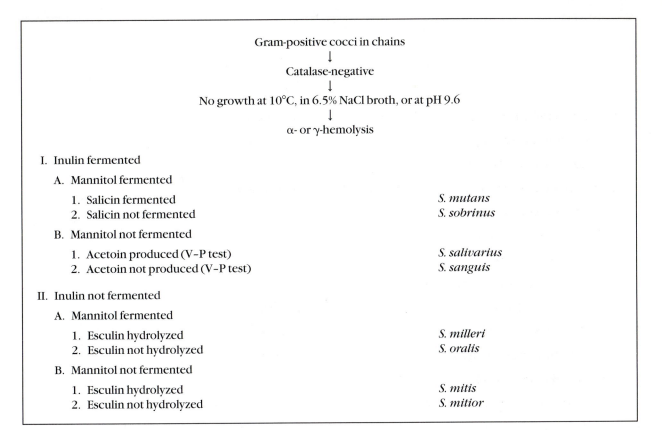

Gram-positive cocci in chains
↓
Catalase-negative
↓
No growth at 10°C, in 6.5% NaCl broth, or at pH 9.6
↓
α- or γ-hemolysis

I. Inulin fermented

    A. Mannitol fermented

        1. Salicin fermented   *S. mutans*
        2. Salicin not fermented   *S. sobrinus*

    B. Mannitol not fermented

        1. Acetoin produced (V–P test)   *S. salivarius*
        2. Acetoin not produced (V–P test)   *S. sanguis*

II. Inulin not fermented

    A. Mannitol fermented

        1. Esculin hydrolyzed   *S. milleri*
        2. Esculin not hydrolyzed   *S. oralis*

    B. Mannitol not fermented

        1. Esculin hydrolyzed   *S. mitis*
        2. Esculin not hydrolyzed   *S. mitior*

Figure H.4

Key to streptococci found in the mouth.

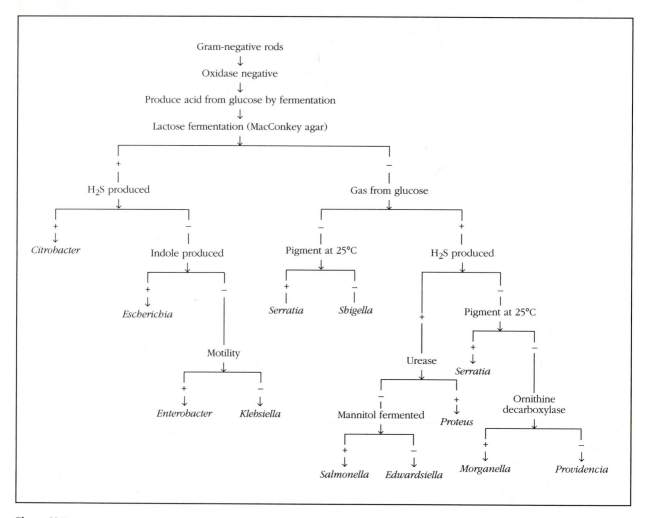

Figure H.5

Identification scheme for enteric genera primarily using MacConkey agar and TSI.

Index